Hamilton-Jacobi equations in Hilbert spaces

V Barbu & G Da Prato

University of Iasi/Scuola Normale Superiore, Pisa

Hamilton-Jacobi equations in Hilbert spaces

π

Pitman Advanced Publishing Program

BOSTON · LONDON · MELBOURNE

PITMAN BOOKS LIMITED
128 Long Acre, London WC2E 9AN

PITMAN PUBLISHING INC
1020 Plain Street, Marshfield, Massachusetts 02050

Associated Companies
Pitman Publishing Pty Ltd, Melbourne
Pitman Publishing New Zealand Ltd, Wellington
Copp Clark Pitman, Toronto

First published 1983

AMS Subject Classifications: 34G, 49C, 60H

Library of Congress Cataloging in Publication Data

Barbu, Viorel.
Hamilton-Jacobi equations in Hilbert spaces.

(Research notes in mathematics; 86)
Includes bibliographical references.
1. Hamilton-Jacobi equations. 2. Initial value problems. 3. Hilbert space. I. Da Prato, Giuseppe. II. Title. III. Series.
QA378.B37 1983 515.3′5 83-4039
ISBN 0-273-08597-2

British Library Cataloguing in Publication Data

Barbu, V.
Hamilton-Jacobi equations in Hilbert spaces.—
(Research notes in mathematics; 86)
1. Hilbert space
I. Title II. Da Prato, G. III. Series
515.7′33 QA322.4

ISBN 0-273-08597-2

Reproduced and printed by photolithography in Great Britain by Biddles Ltd, Guildford

Contents

CHAPTER 1 PRELIMINARIES

1. Basic facts and notations 1
2. Convex functions and generalized gradients 4
3. Evolution equations in Banach spaces 8
4. Control systems in Hilbert spaces 12
5. Stochastic differential equations 14
6. References 18

CHAPTER 2 EXISTENCE IN THE CLASS OF CONVEX FUNCTIONS

1. Introduction 20
2. The contraction semigroup associated with Eq. (1.1) 23
3. Equation (1.1) where A = 0 28
4. Existence and uniqueness for $F(x) = \frac{1}{2}|x|^2$ 35
5. Further regularity results 46
6. An application to construction of optimal feedback controllers 50
7. A variational approach to the Hamilton-Jacobi equations 54
8. Riccati equations 63
9. Stationary Hamilton-Jacobi equations 70
10. References 84

CHAPTER 3 EXISTENCE THEORY IN NONCONVEX CLASSES

1. Introduction 87
2. A direct approach to Hamilton-Jacobi equations in a nonconvex class of solutions 88
3. Applications to synthesis of optimal controls 100
4. A variational approach. Generalized solutions 111
5. Stationary Hamilton-Jacobi equations and control problems with infinite time horizon 127
6. Galerkin approximation of solutions 138
7. References 144

CHAPTER 4 THE DYNAMIC PROGRAMMING EQUATION FOR STOCHASTIC OPTIMAL CONTROL

1. Introduction 145

2. Notations and assumptions 147

3. The linearized problem 149

4. Existence results 153

5. Synthesis of optimal control 163

6. Linear quadratic control 167

7. References 172

Preface

In recent years there has been a growing interest in, and many contributions to, the study of the initial value problem for the Hamilton-Jacobi equations. Major contributions have been given to the existence theory of generalized solutions as well as to the numerical treatment of this problem. The classical Hamilton-Jacobi theory has been enriched with new pertinent results and applications in the modern theory of the calculus of variations, stochastic optimal control and in the foundations of quantum mechanics (the Feynman theory).

In infinite dimensional spaces, particularly in Hilbert spaces, the main motivation to studying Hamilton-Jacobi equations comes from the theory of deterministic distributed parameter systems and control of stochastic distributed systems. In recent years the authors have developed a unifying body of results on existence and approximation of solution to Hamilton-Jacobi equations in Hilbert spaces, and we now feel it appropriate to gather them together in the form of lecture notes. The treatment follows essentially two interrelated methods: a constructive approximating method presented in Chapter 2 and the dynamic programming method.

For the convenience of the reader and to make these lecture notes self-contained, we have summarized in Chapter 1 some basic results in convex analysis and the theory of evolution equations in Banach spaces. The list of references does not pretend to be complete, as it refers only to works directly connected with the subject treated. Part of this work was written when one of the authors was visiting the Scuola Normale Superiore di Pisa (Italy).

V. Barbu, G. Da Prato
Iasi (Romania) and Pisa (Italy)

June 1982

1 Preliminaries

§1 Basic facts and notations

Throughout the following text all Banach spaces are real. The dual of a Banach space X will be denoted by X* and the value of $x^* \in X^*$ on $x \in X$ will be written (x,x^*) or (x^*,x). If X and Y are Banach spaces then $\mathcal{L}(X,Y)$ is the spaces of all continuous linear operators from X to Y endowed with the usual operator norm $\|\cdot\|_{\mathcal{L}(X,Y)}$.

For an operator $A:X \to Y$ we shall denote by D(A) the domain of A, by R(A) the range and by N(A) the null space of A; A^{-1} is the inverse of A* the adjoint of A (if A is linear).

Given a compact interval [a,b] we shall denote by C([a,b];X) the space of all continuous functions from [a,b] to X and by AC ([a,b];X) the space of all absolutely continuous functions on [a,b] with values in X. For a given function $x:[a,b] \to X$ we shall denote by x' the derivative of x in the sense of distributions and we shall denote by $W^{1,p}(a,b;X)$ the space:

$$W^{1,p}(a,b;X) = \{x \in L^p(a,b;X);\ \ x' \in L^p(a,b;X)\} \tag{1.1}$$

where $L^p(a,b;X)$; $1 \leqslant p \leqslant \infty$, is the usual space of all L^p-summable X-valued functions on [a,b]. It is well known (see for instance [4]) that if $x \in W^{1,p}(a,b;X)$ then $x \in AC([a,b];X)$ and

$$\begin{aligned} &\frac{d}{dt}x(t) = x'(t) \quad \text{a.e. } t \in \,]a,b[\\ &x(t) = x(a) + \int_a^t x'(s)ds \text{ for all } t \in [a,b]. \end{aligned} \tag{1.2}$$

For a natural number k we shall denote by $C^{(k)}([a,b];X)$ the space of continuously differentiable functions up to order k, from [a,b] to X.

If [a,b[is an unbounded interval we shall denote by L^p_{loc} (a,b,X) and $W^{1,p}_{loc}$ (a,b;X) the space of functions $x:[a,b] \to X$ having the property that their restrictions to every [a,T], $a < T < b$, belong to $L^p(a,T;X)$ and $W^{1,p}$ (a,T;X) respectively.

Given a bounded and open subset Ω of the Euclidean space $\mathbb{R}^N$ we shall denote by $L^p(\Omega)$ the space of all p-integrable functions on Ω, $1 \leq p \leq \infty$ and by $C^{(k)}(\Omega)$, $C^{(k)}(\bar{\Omega})$, $k = 0,1\ldots$ the usual spaces of k-times continuously differentiable functions on Ω and $\bar{\Omega}$ respectively. By $C_0^{(k)}(\Omega)$ we shall denote the linear space of functions in $C^{(k)}(\Omega)$ with compact supports in Ω. For a natural number k denote by $H^k(\Omega)$ the usual Sobolev space of all functions $f \in L^2(\Omega)$ having all distributional derivatives up to order k in $L^2(\Omega)$. $H_0^k(\Omega)$ is the closure of $C_0^k(\Omega)$ in the norm of $H^k(\Omega)$.

Given a Fréchet differentiable mapping $E:X \to \mathbb{R}$ we shall use either the symbol $\nabla E(x)$ or $E'(x)$, $E_x(x)$, to denote the gradient of E at x. If $E:X \to Y$, by $E^{(k)}$ we shall denote the Fréchet derivative of order $k \geq 1$.

If X and Y are two Banach spaces the following spaces of mappings from X to Y will be used:

1º $C_\infty^k(X;Y)$ is the space of all k-times Fréchet differentiable mappings $E:X \to Y$ such that $E^{(j)}$, $j = 0,1,\ldots,k$ are continuous and bounded on X. The space $C_\infty^k(X;Y)$ is endowed with the norm

$$|E|_{C_\infty^k(X;Y)} = \sum_{j=0}^{k} |E|_j \tag{1.3}$$

where

$$|E|_k = \sup\{|E^{(k)}(x)|;\ x \in X\}. \tag{1.4}$$

(We shall denote by the same symbol $|\cdot|$ the norms in X,Y and in several spaces of linear continuous operators on X.)

2º $C_{Lip,\infty}^k(X;Y)$ is the space of all $E \in C_\infty^k(X;Y)$ such that

$$\|E\|_k = \sup\{|E^{(k)}(x) - E^{(k)}(y)|\ |x - y|^{-1};\ x \neq y, x,y \in X\} < +\infty \tag{1.5}$$

$C_{Lip,\infty}^k(X;Y)$ is a Banach space with the norm:

$$|E|_{C_{Lip,\infty}^k(X;Y)} = \sum_{j=0}^{k} |E|_j + \|E\|_k \tag{1.6}$$

3º For every $r > 0$ denote by Σ_r the closed ball $\{x \in X;\ |x| \leq r\}$ and by $C^k(\Sigma_r,Y)$ the space of all mappings $E:\Sigma_r \to Y$ which are Fréchet differentiable up to order k on Σ_r such that $E^{(j)}$, $j = 0,1,\ldots,k$ are continuous and bounded on Σ_r.

By $C^k_{Lip}(\Sigma_r;Y)$ we shall denote the space of all $E \in C^k(\Sigma_r;Y)$ such that $E^{(k)}$ is Lipschitz on Σ_r, i.e.

$$\|E\|_{k,r} = \sup\{|E^{(k)}(x) - E^{(k)}(y)|\,|x - y|^{-1};\quad x \neq y, x,y \in B_r\} < +\infty. \tag{1.7}$$

The spaces $C^k(\Sigma_r;Y)$ and $C^k_{Lip}(\Sigma_r;Y)$ are endowed with the norms:

$$|E|_{C^k(\Sigma_r;Y)} = \sum_{j=0}^{k} |E|_{j,r} \tag{1.8}$$

$$|E|_{C^k_{Lip}(\Sigma_r;Y)} = \sum_{j=0}^{k} |E|_{j,r} + \|E\|_{k,r} \tag{1.9}$$

where:

$$|E|_{j,r} = \sup\{|E^{(j)}(x)|;\ x \in \Sigma_r\}. \tag{1.10}$$

4º If k is a natural number or zero then $C^k(X;Y)$ is the space of all mappings $E:X \to Y$ which are k-times Fréchet differentiable and their restrictions to every Σ_r belong to $C^k(\Sigma_r;Y)$. The space $C^k_{Lip}(X;Y)$ is defined as the set of all $E \in C^k(X;Y)$ such that the restrictions of E at every Σ_r belong to $C^k_{Lip_0}(\Sigma_r;Y)$.

5º We shall denote by $B([0,T];\ C^k_\infty(X;Y))$ (resp. $B([0,T];\ C^k(X;Y))$) the space of all continuous mappings $E : [0,T] \times X \to Y$ such that

(i) The mapping

$[0,T] \times X \to Y$, $(t,x) \to E^{(j)}(t,x)(z_1,\ldots,Z_j)$ is continuous $j = 1,\ldots,k$, $z_1,\ldots,z_j \in X$

(ii) We have

$$\sup\{|E(t,\cdot)|_j;\ t \in [0,T]\} < +\infty \text{ for } j = 0,1,\ldots,k \tag{1.11}$$

(resp. $\sup\{|E(t,\cdot)|_{j,r};\ t \in [0,T]\} < +\infty$ for $j = 0,1,\ldots,k$, $r > 0\}$).

The space $B([0,T];\ C^k_{Lip,\infty}(X;Y))$ (resp. $B([0,T];\ C^k_{Lip}(X;Y))$ is defined as the set of all $E \in B([0,T]; C^k_\infty(X;Y))$ (resp. $B([0,T]; C^k(X;Y))$ such that:

$$\sup\{\|E(t,\cdot)\|_k ;\ t \in [0,T]\} < +\infty \tag{1.12}$$

(resp. $\sup\{\|E(t,\cdot)\|_{k,r} ;\ t \in [0,T]\} < +\infty \quad \forall r > 0$).

If $Y = \mathbb{R}$ we shall often write $C^k_\infty(X)$ and $C^k_{Lip,\infty}(X)$ instead of $C^k_\infty(X;\mathbb{R})$ and $C^k_{Lip,\infty}(X;\mathbb{R})$ respectively. Also if $k = 0$ we shall simply write $C(X)$, $C_{Lip}(X)$ instead of $C^0(X)$ and $C^0_{Lip}(X)$ respectively.

§2 Convex functions and generalized gradients

Given a real Banach space X and a lower semicontinuous convex function $\phi: X \to \bar{\mathbb{R}} =]-\infty, +\infty]$ we shall denote by $D(\phi) = \{x \in X;\ \phi(x) < +\infty\}$ the effective domain of ϕ and by $\partial\phi: X \to X^*$ the *subdifferential* of ϕ, i.e.

$$\partial\phi(x) = \{x^* \in X^*;\ \phi(x) \leqslant (x^*, x - y) + \phi(y) \text{ for all } y \text{ in } X\}. \tag{2.1}$$

Every element $x^* \in \partial\phi(x)$ is called *subgradient* of ϕ at x. The set of all $x \in X$ for which $\partial\phi(x) \neq \emptyset$ is called the domain of $\partial\phi$ and is denoted $D(\partial\phi)$. If ϕ is Gâteaux differentiable at x then its Gâteaux derivative $\nabla\phi(x)$ coincides with $\partial\phi(x)$. Other properties of $\partial\phi$ are listed in Propositions 1, 2 below (for the proof see ([3])).

PROPOSITION 1 *Let $\phi: X \to \bar{\mathbb{R}}$ be a lower semicontinuous convex function $\not\equiv +\infty$. Then:*

(i) *ϕ is bounded from below by an affine function.*

(ii) *ϕ is continuous on int $D(\phi)$ (the interior of $D(\phi)$).*

(iii) *$D(\partial\phi)$ is a dense subset of $D(\phi)$.*

(iv) *int $D(\phi) \subset D(\partial\phi)$ and $\partial\phi$ is locally bounded on int $D(\phi)$.*

The function $\phi^*: X^* \to \bar{\mathbb{R}}$ defined by

$$\phi^*(x^*) = \sup\{(x,x^*) - \phi(x);\ x \in X\} \tag{2.2}$$

is called the *conjugate* of ϕ.

PROPOSITION 2 *The following conditions are equivalent:*

(j) $x^* \in \partial\phi(x)$

(jj) $x \in \partial\phi^*(x^*)$

(jjj) $\phi(x) + \phi^*(x^*) = (x,x^*)$.

We notice that in the special case when $\phi(x) = \frac{1}{2}|x|^2$ then $\partial\phi = J$ is the *duality mapping* of X, i.e.

$$J(x) = \{x^* \in X^*;\ \ |x|^2 = |x^*|^2 = (x,x^*)\}.$$

A multivalued operator $A:X \to X^*$ is called *monotone* if

$$(x_1-x_2,x_1^*-x_2^*) \geq 0 \text{ for all } x_i \in D(A),\ x_i^* \in Ax_i,\ i = 1,2.$$

The monotone operator A is said to be *maximal monotone* if it has no proper extensions in $X \times X^*$. If $X = H$ is a Hilbert space then according to a well-known result due to Minty (see for instance [1], [4]) the monotone operator A is maximal monotone if and only if $R(\lambda I + A) = H$ for all $\lambda > 0$ (equivalently for some $\lambda > 0$). In this case the operator J_λ defined by

$$J_\lambda = (I + \lambda A)^{-1}$$

is non-expansive and the operator A_λ (the Yosida approximation of A)

$$A_\lambda = \lambda^{-1}(I - J_\lambda) \tag{2.3}$$

is monotone and Lipschitz with Lipschitz constant λ^{-1}. According to a classical result due to R. T. Rockafellar, the subdifferential $\partial\phi$ of a lower semi-continuous convex function ϕ is maximal monotone.

As regards the Yosida approximation A_λ of $A = \partial\phi$ we have (see [4]).

PROPOSITION 3 *Let $X = H$ be a Hilbert space and let $\phi:H \to \bar{\mathbb{R}}$ be a lower semi-continuous convex function $\not\equiv +\infty$. For every $\lambda > 0$ define*

$$\phi_\lambda(x) = \inf\left\{\frac{|x-y|^2}{2\lambda} + \phi(y);\ y \in H\right\},\quad x \in H. \tag{2.4}$$

Then ϕ_λ is convex, Fréchet differentiable and:

$$\phi_\lambda(x) = \frac{\lambda}{2}\,(\partial\phi_\lambda)(x)|^2 + \phi(J_\lambda x),\ \textit{for all } x \in H,\ \lambda > 0 \tag{2.5}$$

$$\lim_{\lambda \to 0} \phi_\lambda(x) = \phi(x) \text{ for all } x \in H \tag{2.6}$$

Given two convex and lower semicontinuous functions ϕ and ψ, in general we cannot say that $\partial(\phi + \psi) = \partial\phi + \partial\psi$. However, we have the following result essentially due to R. T. Rockafellar (see for instance [3] p. 120).

PROPOSITION 4 *Let* $\phi:X \to \bar{\mathbb{R}}$ *and* $\psi:X \to \bar{\mathbb{R}}$ *be two lower semicontinuous and convex functions such that* $(\text{int } D(\phi)) \cap D(\psi) \neq \emptyset$. *Then*

$$\partial(\phi + \psi) = \partial\phi + \partial\psi \tag{2.7}$$

On the same lines we have

PROPOSITION 5 *Let* $X = H$ *be a Hilbert space let* $\phi:H \to \bar{\mathbb{R}}$ $\psi:H \to \bar{\mathbb{R}}$ *be two lower semicontinuous convex functions satisfying the condition,*

$$(\partial\phi(y), \partial\psi_\lambda(y)) \geq -C|\partial\psi_\lambda(y)|(1 + |y|) + C \tag{2.8}$$

for all $y \in D(\partial\phi)$ *and* $\lambda > 0$. *Then* $D(\partial\phi) \cap D(\partial\psi)$ *is a dense subset of* H *and* $\partial(\phi + \psi) = \partial\phi + \partial\psi$.

For the proof see [3] p. 114 (see also [4]).

The following result is known in literature as the Fenchel duality theorem ([3] p. 186).

PROPOSITION 6 *Under assumptions of Proposition 4 we have*

$$\inf \{\phi(x) + \psi(x);\ x \in X\} = \max \{-\phi^*(-x^*) - \psi^*(x^*);\ \ x^* \in X^*\}. \tag{2.9}$$

A monotone operator $A:X \to X^*$ is said to be *coercive* if there exists $x_0 \in X$ such that

$$\lim_{|x| \to \infty} (Ax, x - x_0)/|x| = +\infty.$$

If $A = \partial\phi$ and $\lim_{|x| \to \infty} \phi(x)/|x| = +\infty$ then clearly A is a coercive operator. This happens for instance if the conjugate function ϕ^* is continuous and bounded on every subset of X^* (see [4], [5] p. 100).

We also notice that, according to a well-known result due to F. Brodwer, if

A is maximal monotone and coercive then it is surjective, i.e. $R(A) = X^*$ (see for instance [1] p. 40).

PROPOSITION 7 *Let $\phi:X \to \mathbb{R}$ be a continuous convex function which is bounded on bounded sets. Then $\partial\phi$ is bounded on every bounded subset of X and ϕ is locally Lipschitz.*

Proof For all x,y in X and $|y| = 1$ we have

$$\phi(x) - \phi(x-y) \leq -(\partial\phi(x),y).$$

Hence

$$\sup\{|w ;w \in \partial\phi(x)\} \leq \phi(x-y) - \phi(x) \leq C_r \text{ for } |x| < r.$$

Next, the inequality

$$\phi(x) - \phi(u) \leq (\partial\phi(x),x-u), \text{ for all } x,u \text{ in } X$$

implies that ϕ is locally Lipschitz as claimed. (Here and hereafter throughout by a locally Lipschitz function we mean a function that is Lipschitz on every bounded subset.)

If $\phi':X \times X \to \mathbb{R}$ is the directional derivative of $\phi:X \to \mathbb{R}$, i.e.

$$\phi'(x,h) = \lim_{\lambda \downarrow 0} \lambda^{-1}(\phi(x+\lambda h) - \phi(x))$$

we may define equivalently $\partial\phi$ as:

$$\partial\phi(x) = \{x^* \in X^*;(x^*,h) \leq \phi'(x,h), \text{ for all } h \in X\}. \tag{2.10}$$

Formula (2.10) can be taken for the definition of a more general concept of subgradient due to F. H. Clarke [5], [6].

Let $\phi:X \to \mathbb{R}$ be a locally Lipschitz function on a Banach space X, let x be fixed in X and let h be any point in X. The *generalized directional derivative* of ϕ at x in direction h, denoted $\phi^0(x,h)$, is by definition given by

$$\phi^0(x,h) = \limsup_{\substack{y\to x \\ \lambda\downarrow 0}} \lambda^{-1}(\phi(y + \lambda h) - \phi(y)) . \tag{2.11}$$

The *generalized gradient* of ϕ at x, denoted $\partial\phi(x)$ is the non-empty) set of all

$x^* \in X^*$ satisfying:

$$\phi^0(x,h) \geq (h,x^*) \text{ for all } h \text{ in } X. \tag{2.12}$$

If ϕ is convex then $\phi^0 = \phi'$ and so $\partial\phi$ is just the sub-differential of ϕ in the sense of Definition (2.1). If $X = \mathbb{R}^n$ then $\partial\phi(x)$ is the set $\text{Conv}\{\lim_{n\to\infty} \nabla\phi(x_n);\ x_n \to x\}$. Finally, if ϕ admits a continuous Gâteaux derivative $\nabla\phi$ then $\partial\phi = \nabla\phi$.

In Proposition 8 below we collect for the purpose of Chapter 3 some properties of generalized gradients. For the proof we refer to [5], [6], [16].

PROPOSITION 8 *Let* $\phi : X \to \mathbb{R}$ *be a locally Lipschitz function. Then:*

(i) $\partial\phi(x)$ *is for each* $x \in X$ *a nonempty closed, convex and bounded subset of* X^*.

(ii) $\partial\phi$ *is* w^*-*upper semicontinuous, i.e. if* $x_n^* \in \partial\phi(x_n)$ *where* $x_n \to x$ *strongly in* X *and* $x_n^* \to x^*$ *weak star in* X^* *then* $x^* \in \partial\phi(x)$

(iii) ϕ^0 *is upper semicontinuous on* $X \times X$

(iv) *If* x *and* y *are distinct points of* X *then there is* z *on the line segment between* x *and* y *such that*

$$\phi(y) - \phi(x) \in (y-x),\partial\phi(z)) \tag{2.13}$$

(*The mean value property*.)

§3 Evolution equations in Banach spaces

We begin by reviewing some well-known facts on C_0-semigroups of linear operators. For a detailed treatment of the subject we refer the reader to the classical books of E. Hille and R. S. Phillips [10], K. Yosida [17] and the more recent works [7], [14].

A C_0-semigroup of linear operators on a Banach space X is a map $S(t) : \mathbb{R}^+ = [0, +\infty[\to \mathcal{L}(X,X)$ which satisfies

$$S(t+s) = S(t)S(s) \text{ for all } t,s \geq 0 \tag{3.1}$$

$$S(0) = I \text{ (the identity operator)} \tag{3.2}$$

$$\lim_{t\downarrow 0} S(t)x = x \text{ for all } x \text{ in } X \tag{3.3}$$

The infinitesimal generator A of the semigroup S(t) is defined by

$$Ax = \lim_{t\downarrow 0} t^{-1}(S(t)x-x) \text{ for } x \text{ in } D(A) \tag{3.4}$$

where D(A) is the set of all $x \in X$ for which the limit in (3.4) exists. It turns out that A is a densely defined closed operator. Moreover, for each $x \in D(A)$ the function $t \to S(t)x$ is continuously differentiable on $\mathbb{R}^+$ and satisfies the equation

$$\begin{cases} y'(t) = Ay(t) \text{ for } t \geq 0 \\ y(0) = x \end{cases} \tag{3.5}$$

THEOREM 1 (Hille-Yosida). *A linear, closed and densely defined operator* A *on* X *is the infinitesimal generator of a* C_0*-semigroup on* X *if and only if there exist the real numbers* ω *and* M *such that*

$$|(\lambda I - A)^{-n}|_{\mathcal{L}(X,X)} \leq M(\lambda - \omega)^{-n} \textit{ for all } \lambda > \omega \tag{3.6}$$

and $n \in \mathbb{N}$.

The semigroup S(t) is *contractant*, i.e.

$$|S(t)|_{\mathcal{L}(X,X)} \leq 1 \text{ for all } t \geq 0 \tag{3.7}$$

if and only if A is m-*dissipative*, i.e.

$$|(\lambda I - A)^{-1}|_{\mathcal{L}(X,X)} \leq \lambda^{-1} \text{ for all } \lambda > 0\ . \tag{3.8}$$

The semigroup generated by A is very often denoted e^{tA} or exp(At).

Consider now the inhomogeneous equation.

$$\begin{aligned} &y'(t) = Ay(t) + f(t)\ \ t_0 \leq t \leq T \\ &y(t_0) = x \end{aligned} \tag{3.9}$$

where A is the infinitesimal generator of a C_0-semigroup and $f \in L^1(t_0,T;X)$.

The continuous function $y:[t_0,T] \to X$ defined by the variation of constant formula:

$$y(t) = S(t_0)x + \int_{t_0}^{t} S(t-s)f(s)ds,\ t_0 \leq t \leq T \tag{3.10}$$

is called the 'mild' solution to Eq. (3.9). If $x \in D(A)$ and $f \in C^1([t_0,T];X)$ then $y \in C^1([t_0,T];X)$ satisfies Eq. (3.9) everywhere on $[t_0,T]$.

In general y does not satisfy Eq. (3.9). However, if $x \in D(A)$ and $f \in C([t_0,T];X)$ then:

$$\frac{d^+}{dt} y(t_0) = Ay(t_0) + f(t_0) \tag{3.11}$$

A C_0-semigroup S(t) is said to be *analytic* if for some positive constant C

$$|A\, S(t)|_{\mathcal{L}(X,X)} \leqslant C\, t^{-1} \text{ for } 0 < t \leqslant 1 . \tag{3.12}$$

It turns out that such a semigroup admits a holomorphic extension in a certain sector $\{\lambda \in \mathbb{C}, |\arg \lambda| \leqslant M\}$. For this reason, to characterize the infinitesimal generator of an analytic semigroup it is more convenient to work in the complexified space X.

PROPOSITION 9 *A densely defined linear operator* A *in a complex Banach space* X *is the infinitesimal generator of an analytic semigroup if and only if there exists* $M > 0$ *and* ω *real such that*

$$|(\lambda I - A)^{-1}|_{\mathcal{L}(X,X)} \leqslant M|\lambda - \omega|^{-1} \text{ for } \operatorname{Re} \lambda > \omega. \tag{3.13}$$

It is worth noticing that if S(t) is an analytic semigroup then for every $x \in H$, $S(t)x$ is a continuously differentiable function on $]0, +\infty[$ and satisfies Eq. 3.5 for all $t > 0$. In this context we also notice the following result.

PROPOSITION 10 *Let* X *be a Hilbert space and let* A *be the infinitesimal generator of an analytic semigroup on* X. *Then for each* $f \in L^2(0,T;X)$ *and* $x \in D(A)$ *the 'mild' solution* y *to* Eq. (3.9) *belongs to* $W^{1,2}(0,T;X)$, *satisfies a.e. on* $[0,t]$ Eq. (3.9) *and the estimate*

$$\int_0^T |y'(t)|^2 dt \leqslant C(|Ax|^2 + \int_0^T |f(t)|^2 dt) \tag{3.14}$$

where C *is independent of* x *and* f.

The proof follows by (3.13) via Parseval's formula (see [12]).

A typical example of an infinitesimal generator of an analytic semigroup in X is a linear continuous operator $A \in \mathcal{L}(V,V^*)$ which satisfies the coercivity condition

$$(Ay,y) \leq \alpha|y|_X^2 - \omega|y|_V^2 \text{ for all } y \in V$$

where $\omega > 0$ and V is a real Hilbert space continuously and densely imbedded in X.

Now consider the nonlinear evolution equation

$$\begin{aligned} &y'(t) - Ay(t) \ni f(t) \quad \text{a.e. } t \in]0,T[\\ &y(0) = x \end{aligned} \tag{3.15}$$

in a real Hilbert space H with the norm $|\cdot|$ and scalar product $(\cdot,\cdot)$.

Here $A:H \to H$ is a nonlinear (multivalued) operator of the form $A = -\partial\phi$ where $\phi:H \to \bar{\mathbb{R}}$ is a lower semicontinuous convex function.

PROPOSITION 11 *For every* $x \in \overline{D(A)}$ *and* $f \in L^2(0,T;H)$ *the Cauchy problem* (3.15) *has a unique solution* $y \in C([0,T];H)$ *such that*

$$\sqrt{t}\, y' \in L^2(0,T;H) \tag{3.16}$$

If $x \in D(\phi)$ *then* $y \in W^{1,2}(0,T;H)$ *and if* $x \in D(A)$, $f \in W^{1,2}(0,T;H)$ *then* $y \in W^{1,\infty}(0,T;H)$ *and the right derivative* $d^+/dt\, y(t)$ *exists at all* $t \geq 0$. *Moreover, the map* $(x,f) \to y$ *is nonexpansive from* $H \times L^2(0,T;H)$ *to* $C([0,T];H)$.

The proof of the above result which is due to H. Brézis can be found in [1] p. 188 and in [4].

Given a closed subset C of the Banach space X, we say that the function $S(t):\mathbb{R}^+ \times C \to C$ is a *semigroup of nonlinear contractions* on C if the following conditions hold:

$$S(t+s)x = S(t)S(s)x \text{ for all } t,s \geq 0,\ x \in C \tag{3.17}$$

$$S(0)x = x \text{ for all } x \in C \tag{3.18}$$

$$\lim_{t \downarrow 0} S(t)x = x \text{ for all } x \in C \tag{3.19}$$

$$|S(t)x - S(t)y| \leq |x-y| \text{ for all } t \geq 0 \text{ and } x,y \in C \tag{3.20}$$

The *generator* of the semigroup S(t) is by definition the operator

$$A^0x = \lim_{t \downarrow 0} (S(t)x - x)/t \, . \tag{3.21}$$

It turns out that A^0 is *dissipative* on X, i.e.,

$$|x - y - \lambda(A^0x - A^0y)| \geq |x - y| \tag{3.22}$$

for all $\lambda > 0$ and all x,y in $D(A^0)$.

Conversely, according to a classical result due to Crandall and Ligget (see [1] p. 104) if A is dissipative and

$$R(I - \lambda A) \supset \overline{D(A)} \quad \text{for all } \lambda > 0 \tag{3.23}$$

then A generates a semigroup of nonlinear contractions S(t) on $\overline{D(A)}$ in the following sense

$$S(t)x = \lim_{n \to \infty} \left(I - \frac{t}{n}A\right)^{-n} x, \text{ for } t \geq 0, \, x \in \overline{D(A)} \tag{3.24}$$

If $x \in D(A)$ and X is reflexive then $y(t) = S(t)x$ is absolutely continuous and satisfies the equation

$$\begin{aligned} &y'(t) \in Ay(t) \qquad \text{a.e. } t > 0 \\ &y(0) = x. \end{aligned} \tag{3.25}$$

§4 Control systems in Hilbert spaces

Throughout this section H is a real Hilbert space with the norm $|\cdot|$ and scalar product $(\cdot,\cdot)$.

Consider on H (the state space) the linear evolution process described by the equation

$$\begin{aligned} &y'(t) = Ay(t) + Bu(t), \quad t_0 \leq t \leq T \\ &y(t_0) = y_0 \end{aligned} \tag{4.1}$$

where A is the infinitesimal generator of a C_0-semigroup denoted e^{tA} and $B \in \mathcal{L}(U,H)$ where U is a real Hilbert space with the norm $\|\cdot\|$ and the inner product $\langle\cdot\rangle$ (the *control space*). The function $u:[t_0,T] \to U$ is called *input* or the *control* of the state system (4.1).

Consider the following optimal control problem.

Minimize

$$\int_{t_0}^{T} (g(y(t)) + h(u(t)))dt + \phi_0(y(T)) \tag{4.2}$$

over all $u \in L^2(t_0,T;U)$ and $y \in C([t_0,T];H)$ subject to state system (4.1).

By the solution to (4.1) we mean of course the 'mild' solution, i.e.

$$y(t) = e^{A(t-t_0)} y_0 + \int_{t_0}^{t} e^{A(t-s)}Bu(s)ds, \quad t_0 \leq t \leq T \tag{4.1)'}$$

We shall assume that the following conditions are satisfied.

(a) $g:H \to \mathbb{R}$ and $\phi_0:H \to \mathbb{R}$ are continuous convex functions.

(b) $h:U \to \bar{\mathbb{R}}$ is a lower semicontinuous convex function, non-identically $+\infty$.

A pair (y^*,u^*) for which the infimum in (4.2) is attained is called the *optimal pair*. The control u^* is called *optimal control*. For existence of an optimal control a growth condition must be imposed. As regards the characterization of optimal controls we have

PROPOSITION 12 *Assume that conditions* (a) *and* (b) *are satisfied. If* (y^*,u^*) *is an optimal pair for problem* (4.2) *then there exists a function* $p \in C([t_0,T];H)$ *which satisfies along with* y^* *and* u^* *the system:*

$$(y^*)' = Ay^* + Bu^* \quad t_0 \leq t \leq T \tag{4.3}$$

$$p' \in -A^*p + \partial g(y^*) \quad t_0 \leq t \leq T \tag{4.4}$$

$$B^*p(t) \in \partial h(u^*(t)) \quad \text{a.e. } t \in]t_0,T[\tag{4.5}$$

$$p(T) + \partial\phi_0(y^*(T)); y^*(0) = y_0. \tag{4.6}$$

Proposition 12 has been proved in [2] (see also [3] p. 269) in a more general context.

The arc p which occurs in Proposition 12 is called *dual extremal* associated with the optimal pair (y^*,u^*).

By definition, a *feedback control* for system (4.1) is a function (possibly multivalued) $\Lambda:[t_0,T] \times H \to U$ having the property that for each $y_0 \in H$ and $s \in [t_0,T]$ the Cauchy problem

$$\begin{cases} y' = Ay + B\Lambda(t,y), & s < t < T \\ y(s) = y_0 \end{cases} \tag{4.7}$$

has a solution $y(t,s,y_0)$. A feedback control Λ is called *optimal feedback control* if for all $s \in [t_0,T]$ and $y_0 \in H$, $u(t) = \Lambda(t,y(t,s,y_0))$ is an optimal control of Problem (4.2) on the interval $[s,T]$. The function $\phi:[t,T] \times H \to \mathbb{R}$ defined by

$$\phi(t,y_0) = \inf \{\int_t^T (g(y(s,t,y_0)) + h(u(s)))ds + \phi_0(y(T,t,y_0)); \; u \in L^2(t,T;U)\} \tag{4.8}$$

is called the optimal value function of Problem (4.2).

In the special case where

$$g(y) = \frac{1}{2}(Qy,y), \quad h(u) = \frac{1}{2}\langle Nu,u\rangle \tag{4.9}$$

$$\phi_0(y) = \frac{1}{2}(Ey,y) \text{ for } y \in H \tag{4.10}$$

the optimality system (4.3) $\sim$ (4.6) becomes

$$\begin{aligned} y^{*\prime} &= Ay^* + Bu^* \quad 0 < t < T \\ p' &= -A^*p + Qy^* \quad 0 < t < T \end{aligned} \tag{4.11}$$

$$B^*p = Nu^* \qquad \text{a.e. } t \in]0,T[\tag{4.12}$$

$$y^*(0) = y_0, \; p(T) = -Ey^*(T). \tag{4.13}$$

For a detailed treatment of quadratic control problems we refer the reader to Lions' book [11] (see also [7]).

§5 Stochastic differential equations

Let (Ω,ε,P) be a probability space, H and K separable Hilbert spaces. A mapping $X:[0,T] \times \Omega \to H$ such that $X(t,\cdot)$ is ε-measurable $\forall\, t \in [0,T]$ is called a H-*valued stochastic process*. We shall write $X(t,\omega) = X_t(\omega)$ for any $t \in [0,T]$ and $\omega \in \Omega$.

The process X is called *measurable* if it is $B_{[0,T]} \times \varepsilon$-measurable, where $B_{[0,T]}$ is the σ-algebra of Borelian sets on $[0,T]$; X is called *pathwise continuous* or *continuous* if $X(\cdot,\omega)$ is continuous in $[0,T]$ for almost $\omega \in \Omega$.

A *H-valued Brownian motion* is a H-valued stochastic process W such that for any $t > s \geq 0$ we have:

(i) $W_t - W_s$ is independent of $W_s - W_0$ for any $t > s > \sigma > 0$.

(ii) $W_t - W_s$ has a multinormal density with covariance S a nuclear symmetric operator, namely

$$\text{cov } (W_t - W_s) = (t-s)S. \tag{5.1}$$

(iii) $W_0 = 0$.

It is possible to represent a H-valued Brownian motion in the following form (see for instance [7], [13]):

$$W(t) = \sum_{k=1}^{\infty} \sqrt{\lambda}_k e_k \, W_k(t) \tag{5.2}$$

where $\{e_k\}$ is an orthonormal basis in H, $\lambda_k \geq 0$ and $\{W_k\}$ is a sequence of mutually independent real Brownian motions; in this case we have

$$S\,x = \sum_{k=1}^{\infty} \lambda_k (x,e_k)_H \, e_k \text{ for all } x \in H. \tag{5.3}$$

We shall denote by F_t^W the σ-algebra generated by $\{W(s), s \leq t\}$, and by $L_W^2(0,T;H)$ the set of all the H-valued measurable processes X such that

(i) $\int_0^T |X(s,\omega)|^2 ds < +\infty$ ω a.e. in Ω.

(ii) X_t is F_t^W-measurable.

Finally we set

$$M_W^2\,(0,T;H) = \{X \in L_W^2(0,T,H);\ E \int_0^T |X(s)|^2 ds < +\infty\} \tag{5.4}$$

where E means expectation.

We shall denote by $L_W^2(0,T;\mathcal{L}(K;H))$ the set of all the $\mathcal{L}(K;H)$-valued mappings X such that:

(a) X is weakly measurable with respect to the σ-algebra $B_{[0,T]} \times \varepsilon$.

(b) $\int_0^T |X(s,\omega)|^2 ds < +\infty \quad \omega$ a.e. in Ω.

(c) $X(t,\cdot)$ is F_t^W-measurable.

Moreover we set:

$$M_W^2(0,T;\mathcal{L}(K;H)) = \{X \in L_W^2(0,T;\mathcal{L}(K;H)); \quad E\int_0^T |X(s)|^2 ds < +\infty\}. \tag{5.5}$$

For each $X \in L_W^2(0,T;\mathcal{L}(K;H))$ we can define the Itô integral (see for instance [13])

$$I(t) = \int_0^T X(s)dW_s. \tag{5.6}$$

It can be shown that I is a pathwise continuous process, moreover if $X \in M_W^2(0,T;H)$ we have

$$E\int_0^T X(s)dW_s = 0 \tag{5.7}$$

$$E|\int_0^T X(s)dW_s|^2 = E\int_0^T \mathrm{Tr}(X(s)SX^*(s))ds \tag{5.8}$$

where $\mathrm{Tr}(\Sigma)$ means the *trace* of the linear operator Σ.

PROPOSITION 13 *Let* $X \in L_W^2(0,T;\mathcal{L}(K;H))$; *assume that for* $p \geqslant 2$ *we have*

$$E|\int_0^T X(s)dW_s|^p < +\infty. \tag{5.9}$$

Then there exists $C_p > 0$ *such that*

$$E|\int_0^T X(s)dW_s|^p \leqslant C_p T^{\frac{p}{2}-1} E\int_0^T |X(s)|^p ds\ . \tag{5.10}$$

For the proof see [13].

Let X be a K-valued stochastic process such that

$$X(t) = X(s) + \int_s^t b(r)dr + \int_s^t \sigma(r)dW_r,\ t > s \tag{5.11}$$

where $b \in L^2_W(0,T;K)$ and $\sigma \in L^2_W(0,T;\mathcal{L}(K;H))$.

Then we write

$$dX = bdt + \sigma dW_t \tag{5.12}$$

and call dX the *stochastic differential* of X.

PROPOSITION 14 (Itô formula)

Let $\phi \in C^1([0,T] \times K;\mathbb{R})$ *such that* $\phi_{xx} \in C([0,T];\mathcal{L}(K,K))$. *Then if* $dX = bdt + \sigma dW$ *we have:*

$$\begin{aligned} d\phi(t,X) = \phi_t(t,X)dt + (\phi_x(t,X),dX) \\ + \frac{1}{2} \operatorname{Tr}(\sigma S\sigma^* \phi_{xx}(t,X)). \end{aligned} \tag{5.13}$$

Consider finally the following stochastic differential equation:

$$\begin{aligned} dy &= (Ay + f(t,y))dt + g(t,y)dW_t \\ y(0) &= y_0 \in K \end{aligned} \tag{5.14}$$

where $A:D(A) \subset K \to K$ is the infinitesimal generator of a C_0-semigroup in K, $f \in C([0,T] \times K;K)$, $g \in C([0,T] \quad K;\mathcal{L}(K;H))$.

The process $y \in L^2_W(0,T;K)$ is called a *mild solution* to Eq. (5.14) if:

$$\begin{aligned} y(t) = e^{tA}y_0 + \int_0^t e^{(t-s)A}f(s,y(s))ds \\ + \int_0^t e^{(t-s)A}g(s,y(s))dW_s \,. \end{aligned} \tag{5.15}$$

The following proposition is proved in [8].

PROPOSITION 15 *Assume that:*

(a) f *is locally Lipschitz uniformly in* t

(b) g *is Lipschitz uniformly in* t

(c) *There exist* L, N $\in \mathbb{R}$ *such that*

$$(Ax,x) + (f(t,x),x) \leq L + N|x|^2 \text{ for all } x \in K. \tag{5.16}$$

Then there exists a unique mild solution to Eq. (5.14) $y \in M_W^2(0,T;K)$.

§6 References

[1] V. Barbu, *Nonlinear semigroups and differential equations in Banach spaces*, Noordhoff, Leyden, 1976.

[2] V. Barbu, Constrained control problems with convex cost in Hilbert spaces, *J. Math. Anal. Appl.*, *55* (1976), 502-528.

[3] V. Barbu and Th. Precupanu, *Convexity and optimization in Banach spaces*, Sijthoff and Noordhoff, International Publishers, Alphen aan den Rijn, 1978.

[4] H. Brézis, Opérateurs maximaux monotones et semigroupes de contractions dans les espaces de Hilbert, Math. Studies 5, North-Holland, Amsterdam, 1973.

[5] F. Clarke, Generalized gradients and applications, *Trans. Amer. Math. Soc.*, *205* (1975), 247-262.

[6] F. Clarke, Generalized gradients of Lipschitz functionals, *Advances in Math.*, *40* (1981), 52-67.

[7] R. Curtain and A.J. Pritchard, *Infinite dimensional linear systems theory*, Lecture notes in control and information sciences 8 Springer-Verlag, Wiesbaden, 1978.

[8] G. Da Prato, M. Iannelli and L. Tubaro, Semilinear stochastic differential equations in Hilbert spaces, *Bollettino Unione Matematica Italiana*, (5) *16* - A (1978), 168-185.

[9] A. Friedman, *Stochastic differential equations and applications*, Academic Press, New York, 1975.

[10] E. Hille and R.S. Phillips, *Functional analysis and semigroups*, American Mathematical Society,Providence, R.I., 1957.

[11] J.L. Lions, *Optimal control of systems described by partial differential equations*, Springer-Verlag, Wiesbaden, 1971.

[12] J.L. Lions and E. Magenes, Problèmes aux limites non homogènes et applications, Dunod, Gauthier-Villars, Paris 1970.

[13] M. Metivier and T. Pellaumail, *Stochastic integral*, Academic Press, 1980.

[14] A. Pazy, *Semigroups of linear operators and applications to partial differential equations*, Lecture Notes 10, Maryland, 1974.

[15] R.T. Rockafellar, *Convex analysis*, Princeton University Press, Princeton, 1970.

[16] R.T. Rockafellar, Directionally Lipschitzian functions and subdifferential calculus, *Proc. London Math. Soc. 39* (1979), 331-355.

[17] K. Yosida, *Functional analysis*, Springer-Verlag, Wiesbaden, 1965.

2 Existence in the class of convex functions

§1 Introduction

In this chapter we shall discuss the existence and uniqueness of solution for the Hamilton-Jacobi equations

$$\phi_t(t,x) + F(\phi_x(t,x)) - (Ax,\phi_x(t,x)) = g(t,x);t \in [0,T]\ x \in D(A) \tag{1.1}$$

with the initial value condition

$$\phi(0,x) = \phi_0(x) \qquad x \in H \tag{1.2}$$

in the class of convex functions on a real Hilbert space H.

The solution ϕ is a function on $[0,T] \times H$ with values in H and the subscripts t and x denote the partial derivatives with respect to t and x. The operator $A:D(A) \subset H \to H$ is the infinitesimal generator of a C_0-semigroup of linear continuous operators on H and $F:H \to \mathbb{R}$, $\phi_0:H \to \mathbb{R}$, $g:[0,T] \times H \to \mathbb{R}$ are given convex functions (in x).

There exists an extensive literature on the Hamilton-Jacobi equations in $\mathbb{R}^N$. In this context we cite the works of S. Kružkov [24], A. Douglis [19] and W. Fleming [20] where the existence and uniqueness of a global solution is established in the class of semi-concave functions on $\mathbb{R}^N$ under the main assumption that F is a convex and continuously differentiable function. The Kružkov approach to the problem consists in approximating Eq. (1.1) where $A \equiv 0$, $g \equiv 0$ by the family of semilinear parabolic equations

$$\begin{aligned} &\phi_t + F(\phi_x) - \varepsilon\Delta\phi = 0 \quad [0,T] \times \mathbb{R}^N \\ &\phi(0,x) = \phi_0(x) \end{aligned} \tag{1.3}$$

and after letting ε tend to zero. A. Douglis [19] has used a difference scheme method and W. Fleming [20] a variational approach.

More recently in the existence theory for Eq. (1.1) a new technique has been developed which relies on the theory of nonlinear semigroups of contractions ([2] [10] [11]). The idea, first used by M.G. Crandall [14] in the study of the hyperbolic conservation law equation, consists in writing Eq. (1.1) with the initial value condition (1.2) as a nonlinear evolution equation of the form

$$\begin{aligned} &z'(t) = Az(t) \qquad 0 \leq t \leq T \\ &z(0) = \phi_0 \end{aligned} \tag{1.4}$$

in the space X of all bounded, uniformly continuous functions on $\mathbb{R}^N$. The operator $A:X \to X$ is defined as follows. For each $k > 0$ define

$$D_k = \{y \in L^\infty(\mathbb{R}^N);\ \|y\|_{L^\infty(\mathbb{R}^N)} \leq k,\ |y(x+h)-y(x)| \leq k|h|,$$

$y(x+h)-2y(x)+y(x-h) \leq k|h|^2$ for all $h,x \in \mathbb{R}^N\}$. We set $D_k^0 = \{y \in D_x;\ \exists\ \lambda_y > 0,$ $y - \lambda_y F(y_x) \in D_k\}$ and define A as the closure in $L^\infty(\mathbb{R}^N)$ of the operator A_∞,

$$A_\infty y = -F(y_x) \text{ for } y \in \bigcup_{k>0} D_k^0 . \tag{1.5}$$

It turns out that if F is convex and belongs to $C^2(\mathbb{R}^N)$ then $\overline{D(A)} = X$, A is dissipative and $R(I - \lambda A) = X$ for $\lambda > 0$. Then in virtue of the Crandall-Ligget theorem (see §3 in Chapter 1), A generates a semigroup of contractions S(t) on X. Thus the Cauchy Problem (1.4) has a unique generalized solution $z(t) = S(t)\phi_0$. Furthermore, if $\phi_0 \in D_k$ for some k then $z(t) = \phi(t,x)$ is locally Lipschitz in $\mathbb{R}^+ \times \mathbb{R}^N$ and satisfies a.e. Eq. (1.1) (see [19], [11]). We may therefore conclude that *if* $F \in C^2(\mathbb{R}^N)$ *is convex and if* $\phi_0 \in D_k$ *then there exists a unique solution* ϕ *to Problem (1.1), (1.2)* ($A \equiv 0$, $g \equiv 0$) *such that* $\phi(t) \in D_k$ *for all* $t \geq 0$.

We have mentioned this typical existence and uniqueness result in $\mathbb{R}^N$ (recently, M.G. Crandall and J.P. Lions [15] have obtained a more general result which does not assume the convexity of F) in order to be able to compare the results of this chapter and the next with the existing literature on classical Hamilton-Jacobi equations (see [12] and [22] for significant results and complete references).

The present approach is quite different from that used in the above mentioned works though it still exploits the accretivity of the operator $\phi \to F(\phi_x)$ in an appropriate function space. Briefly, the idea consists in approximating $F(\phi_x)$ by

$$\varepsilon^{-1}(\phi - (\phi^* + \varepsilon F)^*)$$

(ϕ^* is the conjugate of ϕ) and letting ε tend to zero in the approximating equations. The contents of the chapter are outlined below.

In §2 the operator $\phi \to - F(\phi_x)$ is studied and it is shown that it is the infinitesimal generator of semigroups of contractions on $C(H)$. In §3 Eq. (1.1) is studied where $A \equiv 0$ and an explicit form of the solution ϕ is given. In §4, §5 Eq. (1.1) is studied where $F(y) = \frac{1}{2}|y|^2$. There is a close connection between Eq. (1.1) and the operator equation

$$E_t - E_x A - A^*E + E_x E = G, \; E(0,x) = E_0(x). \tag{1.6}$$

Namely, if ϕ satisfies Eq. (1.1) where $F_y = \frac{1}{2}|y|^2$ then $E = \phi_x$ satisfies Eq. (1.6) where $G = \phi_x$ and $E_0 = (\phi_0)_x$. In the special case where G and E_0 are linear operators, Eq. (1.6) reduces to the well-known Riccati equation

$$P'(t) - P(t)A - A^*P(t) + P^2(t) = Q; \; P(0) = P_0 \tag{1.7}$$

which will be studied by some specific methods in §8.

There exists another classical method for the study of Hamilton-Jacobi equations which roughly speaking consists in looking for the solution ϕ to Eq. (1.1) as the optimal value function (the action function is classical mechanics) of a certain problem of the calculus of variations. This approach is illustrated in §7. Conversely every existence result for Eq. (1.1) leads to an optimal feedback controller for a control problem associated with a state system of the form

$$x' = Ax + Bu; \; x(0) = x_0.$$

This problem is presented in §6. §9 deals with the stationary Hamilton-Jacobi equation

$$\frac{1}{2}|\phi_x|^2 - (Ax, \phi_x) = g \tag{1.8}$$

in the class of convex continuous function ϕ on H.

The main part of the material presented here relies on the authors' joint works [6], [7], [8], [9].

§2 The contraction semigroup associated with Eq. (1.1)

First let us set forth the assumptions which will be in effect throughout this chapter:

(i) H *is a real Hilbert space with inner product* (,) *and norm* | |.

(ii) $F \in C^1_{Lip}(H)$ *is convex*, $F'(0) = 0$ *and*

$$\lim_{|x|\to\infty} F(x)/|x| = +\infty. \tag{2.1}$$

(iii) *For every* $r > 0$ *there exists* $\omega_r > 0$ *such that:*

$$(F'(x) - F'(y), x - y) \geq \omega_r |x-y|^2 \quad \forall x,y \in \Sigma_r. \tag{2.2}$$

(iv) *The linear operator* A *is the infinitesimal generator of a* C_0*-semigroup of contractions* e^{tA} *on* H.

We shall denote by K the set of all convex functions $\phi \in C(H)$ satisfying the conditions:

$$(F'(y), x) \geq 0, \quad \forall x \in D(\partial\phi),\ y \in \partial\phi(x);\ 0 \in \partial\phi(0). \tag{2.3}$$

We begin with the following technical lemma.

LEMMA 1. *Let* F *satisfy conditions* (i) - (iii) *and let* $\phi \in K$. *Then for every* $t > 0$ *the equation:*

$$y + t\, F'\, \partial\phi(y) \ni x \quad \forall x \in H \tag{2.4}$$

has a unique solution $y = y_t(x)$ *which satisfies*

$$|y_t(x)| \leq |x| \quad \forall x \in H \tag{2.5}$$

Moreover, for each $t > 0$ *the map* $x \to y_t(x)$ *belong to* $C_{Lip}(H,H)$ *and*

$$|x - y_t(x)| \leq t\, |F|_{1,\delta(\phi,r)} \text{ for all } x \in \Sigma_r \tag{2.6}$$

where $\delta(\phi,r) = \sup\{|z|; z \in \partial\phi(y);\ y \in \Sigma_r\}$. *If* $\phi \in C^1(H)$ *then*

$$\phi'(y_t(x)) = (F')^{-1}(t^{-1}(x - y_t(x))). \tag{2.7}$$

Proof By assumption (2.2) the mapping $(F)^{-1}$ is one-to-one and Eq. (2.4) is equivalent to

$$\partial\phi(y) - (F')^{-1}(t^{-1}(x - y)) \ni 0. \tag{2.8}$$

Moreover, since $F' \in C(H,H)$ we may infer that $(F')^{-1}$ is coercive on H and according to Proposition 4 in Chapter 1 and the discussion preceding it the operator $\Gamma y = \partial\phi(y) - (F')^{-1}(t^{-1}(x - y))$ is maximal monotone and surjective. Hence Eq. (2.4) has at least one solution $y = y_t(x)$. By condition (2.3) we see that y_t satisfies (2.5).

Now let $x,\tilde{x} \in \Sigma_r$ and $u \in \partial\phi(y)$, $\tilde{u} \in \partial\phi(\tilde{y})$ be such that

$$y + t\,F'(u) = x,\ \tilde{y} + t\,F'(\tilde{u}) = \tilde{x}. \tag{2.9}$$

We have

$$t(F'(u) - F'(\tilde{u}),\ u - \tilde{u}) \leq (u - \tilde{u}, x - \tilde{x})$$

and by (2.2) we have

$$|u - \tilde{u}| \leq (t\omega_R)^{-1}|x - \tilde{x}|. \tag{2.10}$$

Hence the map $x \to y_t(x)$ is locally Lipschitz on H. On the other hand, by (2.8) it follows that

$$|x - y_t(x)| \leq t \sup \{|F'(z)|; z \in \partial\phi(y_t(x))\}.$$

Since ϕ is bounded on bounded subsets so is $\partial\phi$ (see Proposition 6 in Chapter 1). Thus the latter implies (2.6). Now assume that $\phi \in C^1(H)$. Then Eq. (2.8) can be equivalently written in the form (2.7). The proof is complete.

For every $\phi \in K$ and $t \geq 0$ define:

$$(S(t)\phi)(x) = (\phi^* + tF)^*(x) \quad \forall x \in H \tag{2.11}$$

where * is the symbol for the conjugate (see §2, formula (2.1) in Chapter 1). Inasmuch as

$$\lim_{|x|\to\infty} (\phi^*(x) + tF(x))/|x| = +\infty \text{ for } t > 0 \tag{2.12}$$

it follows that for each $t > 0$, $S(t)\phi$ is a continuous convex function on H.

Moreover, by the Fenchel duality theorem (Proposition 6 in Chapter 1) $S(t)\phi$ can be equivalently defined (for $t > 0$) as:

$$(S(t)\phi)(x) = \inf \{\phi(y) + tF^*(t^{-1}(x-y));y \in H\}$$

$$= \phi(y_t(x)) + tF^*(t^{-1}(x-y_t(x)))$$

$$= \phi(y_t(x)) + tF^*F' \partial\phi(y_t(x)), \quad \forall x \in H \tag{2.13}$$

It turns out that $S(t)$ is a semigroup of nonlinear contractions on $C(H)$ endowed with the family of seminorms $\{|\cdot|_{0,r}\}_{r>0}$, where

$$|\phi|_{0,r} = \sup \{|\phi(x) ; x \in \Sigma_r\}.$$

LEMMA 2. *We have*

$$S(t)(K) \subset K \text{ for all } t \geq 0. \tag{2.14}$$

$$S(t + s)\phi = S(t)S(s)\phi \text{ for all } t,s \geq 0 \text{ and } \phi \in K. \tag{2.15}$$

$$|S(t)\phi - S(t)\bar{\phi}|_{0,r} \leq |\phi - \bar{\phi}|_{0,r} \text{ for all } \phi, \bar{\phi} \in K \text{ and } r > 0. \tag{2.16}$$

$$\lim_{t\to 0} S(t)\phi = \phi \text{ in } C(H) \text{ for all } \phi \in K. \tag{2.17}$$

Moreover, for every $t > 0$, $S(t)$ *maps* K *into* $C^1(H)$ *and*

$$\lim_{t\to 0} t^{-1}(\phi - S(t)\phi) = F(\phi') \text{ in } C(H) \tag{2.18}$$

for each $\phi \in K \cap C^1(H)$ *having the property that* ϕ' *is uniformly continuous on every bounded subset of* H.

Proof Let $\phi,\bar{\phi}$ be two elements of K. Since in virtue of Lemma 1, $y_t(x),\bar{y}_t(x)$ ($\bar{y}_t$ is the solution to (2.4) relative to $\bar{\phi}$) belong to Σ_r whenever $|x| \leq r$ it follows by (2.13) that

$$(S(t)\bar{\phi})(x) - (S(t)\phi)(x) = \min \{\bar{\phi}(y) + tF^*\left(\frac{x-y}{t}\right), y \in H\}$$

$$- \phi(y_t(x)) - tF^* \left(\frac{x-y_t(x)}{t}\right) \leq \bar{\phi}(y_t(x)) - \phi(y_t(x))$$

$$\leq |\phi-\bar{\phi}|_{0,r}$$

and (2.16) is fulfilled. We have also proved that $S(t)\phi$ is bounded on bounded subsets. Next by (2.11) and Proposition 2 (Chapter 1) we see that:

$$\partial(S(t)\phi)(x) = (\partial\phi^* + tF')^{-1}(x) \quad \forall x \in H,\ t > 0. \tag{2.19}$$

Now by the above cited surjectivity result the operator $\partial\phi^* + tF'$ is maximal monotone and by condition (2.2) it is coercive. We may infer that for $t > 0$ the mapping $(\partial\phi^* + tF')^{-1}$ is single valued and Lipschitzian on every bounded subset of H. This implies that $S(t)\phi \in C^1(H)$ for $t > 0$ and by (2.6), (2.19) we see that:

$$\begin{aligned}(S(t)\phi)_x(x) &= (\partial\phi^* + tF')^{-1}(x)\\ &= (F')^{-1}(t^{-1}(x-y_t(x))) \text{ for } x \in H,\ t > 0.\end{aligned} \tag{2.20}$$

In particular, we have for all $t > 0$ and $x \in H$,

$$(F'((S(t)\phi)_x(x),x) = t^{-1}(x-y_t(x),x) \geq 0$$

while by Lemma 1 and (2.20), $(S(t)\phi)_x(0) = 0$.

We have therefore shown that $S(t)\phi \in K$ for every $t > 0$ as claimed. To prove the semigroup property (2.15) we use again the Fenchel theorem in (2.11) and (2.13) to get

$$\begin{aligned}(S(t + s)\phi)(x) &= \inf \{(\phi^* + sF)^*(y) + tF^*(t^{-1}(x-y));y \in H\}\\ &= \inf \{(S(s)\phi)(y) + tF^*(t^{-1}(x-y);y \in H\}\\ &= (S(t)S(s))\phi)(x)\end{aligned}$$

as claimed.

Recall now the conjugacy formula (see Proposition 2 in Chapter 1)

$$F^*(F'(z)) + F(z) = (F'(z),z) \quad \forall z \in H \tag{2.21}$$

which implies $F^*(F'(z)) \geq 0 \quad \forall z \in H$; by (2.13) it follows

$$\phi(y_t(x)) \leq (S(t)\phi)(x) \leq \phi(x) \tag{2.22}$$

which along with (2.6) proves (2.17) if ϕ belongs to $C^1(H)$ the general case follows from the density of $C^1(H) \cap K$ in K (see Chapter 1). It remains to prove (2.18). We fix $\phi \in K \cap C^1(H)$ and notice that by (2.13) we have:

$$\phi(x) - (S(t)\phi)(x) = \phi(x) - \phi(y_t(x)) - tF^*(t^{-1}(x-y_t(x)))$$

$$\leq (\phi'(x),x-y_t(x)) - tF^*(t^{-1}(x-y_t(x)))$$

$$= (\phi'(x),x-y_t(x)) - tF^*F'\phi'(y_t(x))$$

by virtue of (2.4). Using (2.21) it follows that

$$t^{-1}(\phi(x) - (S(t)\phi)(x)) \leq (\phi'(x) - \phi'(y_t(x)),F'\phi'(y_tx)) + F\phi'(y_t(x)). \tag{2.23}$$

On the other hand, since F is convex,

$$F(\phi'(y_t(x)) \leq F\phi'(x) + (F'\phi'(y_t(x)),\phi'(y_t(x)) - \phi'(x))$$

which along with (2.23) yields

$$t^{-1}(\phi(x) - (S(t)\phi)(x)) - F(\phi'(x)) \leq 0 \tag{2.24}$$

Similarly, by (2.13) and (2.23) it follows that

$$(S(t)\phi)(x) - \phi(x) \leq (\phi'(y_t(x)),y_t(x) - x)$$

$$+ tF^*(t^{-1}(x-y_t(x))) = - t(\phi'(y_t(x)),F'(\phi'(y_t(x)))$$

$$+ tF^*(F'(\phi'(y_t(x))) = - tF(\phi'(y_t(x)))$$

and therefore

$$t^{-1}(\phi(x) - (S(t)\phi)(x)) - F(\phi'(y_tx))) \geq 0. \tag{2.25}$$

By (2.24), (2.25) we infer that

$$0 \leq F(\phi'(x)) - t^{-1}(\phi(x) - (S(t)\phi)(x))$$

$$\leq F(\phi'(x)) - F(\phi'(y_t(x))) \tag{2.26}$$

$$\leq (F'(\phi'(x)),\phi'(x) - \phi'(y_t(x))); \; t > 0, \; x \in H.$$

Since ϕ' is uniformly continuous on Σ_r it follows by (2.21) and (2.26) equality (2.18) thereby completing the proof of Lemma 2.

REMARK 1. Let $C^1_u(H)$ be the space of all $\phi_0 \in C^1(H)$ such that ϕ_0' is uniformly continuous on bounded subsets. Let L_0 be the operator on C(H) defined by

$$L_0(\phi) = - F(\phi') \text{ for } \phi \in D(L_0) = C^1_u(H). \tag{2.27}$$

If $\phi_i \in D(L_0)$, i = 1,2, and x_i are defined by

$$\phi_i + \alpha\left(\frac{\phi_i - S(h)\phi_i}{h}\right) = x_i \qquad \alpha > 0$$

we have, in virtue of (2.16)

$$|\phi_1 - \phi_2|_{0,r} \leq |x_1 - x_2|_{0,r} = \left|\phi_1 - \phi_2 + \alpha\left(\frac{\phi_1 - S(h)\phi_1}{h}\right) - \alpha\left(\frac{\phi_2 - S(h)\phi_2}{h}\right)\right|_{0,r}$$

and for $h \to 0^+$ by Lemma 2 follows:

$$|\phi_1 - \phi_2|_{0,r} \leq |\phi_1 - \phi_2 - \alpha(L_0\phi_1 - L_0\phi_2)|_{0,r} \qquad \forall\alpha > 0, \quad \forall r > 0 \tag{2.28}$$

that is L_0 is 'dissipative'. According to a well-known result of Kato the dissipativity condition (2.28) admits an equivalent representation (see [3] page 71).

$$(L_0\phi_1 - L_0\phi_2, f) \leq 0 \text{ for some } f \in \Phi_r(\phi_1 - \phi_2) \tag{2.29}$$

where $\Phi_r : C(\Sigma_r;\mathbb{R}) \to (C(\Sigma_r;\mathbb{R}))^*$ is the duality mapping of the space $C(\Sigma_r;\mathbb{R})$.

§3 Equation (1.1) where A = 0

To begin with we consider the homogeneous Cauchy problem:

$$\begin{aligned} &\phi_t(t,x) + F(\phi_x(t,x)) = 0, \ t \geq 0, \ x \in H \\ &\phi(0,x) = \phi_0(x), \ x \in H \end{aligned} \tag{3.1}$$

where F satisfies assumptions (ii), (iii) and $\phi_0 \in K$.

If $S(t):\mathbb{R}^+ \times K \to K$ is the semigroup defined by formula (2.11) we define the function $\phi:\mathbb{R}^+ \times H \to \mathbb{R}$,

$$\phi(t,x) = (S(t)\phi_0)(x);\ t \geqslant 0,\ x \in H\ . \tag{3.2}$$

By Lemma 2 it follows that $\phi(t,\cdot) \in C^1(H)$ for every $t > 0$ and $\phi(t) \in K$ for all $t \geqslant 0$. Moreover, by (2.15) and (2.18) it follows that for each $\phi_0 \in K$ the right derivative $d^+/dt\ \phi(t,x)$ as a function from $[0,T]$ to $C(H)$ exists at every $t > 0$ and

$$\frac{d^+}{dt}\phi(t,x) + F(\phi_x(t,x)) = 0 \quad \forall t > 0 \tag{3.3}$$

(because by (2.20) $(S(t)\phi_0)_x$ is uniformly continuous on bounded subsets). Moreover, if $\phi_0 \in C_u^1(H;\mathbb{R})$ then by Lemma 2, Eq. (3.3) is also satisfied at $t = 0$.

It is worth noticing that by (2.20) and (2.7),

$$\phi \in C(]0,T];\ C^1(H))\ (\phi \in C([0,T];\ C^1(H))\text{ if }\phi_0 \in C_u^1(H))\quad\text{and}$$

$$\sup\{|\phi(t,\cdot)|_{1,r};\ 0 < t \leqslant T\} \leqslant C_r \quad \forall r > 0\ . \tag{3.4}$$

Hence the function $t \to \frac{d^+}{dt}\phi(t,\cdot)$ is continuous from $]0,T[$ to $C(H)$ (from $[0,T[$ to $C(H)$ if $\phi_0 \in C_u^1(H)$). By a standard argument (see for instance [1] page 238) it follows that $\phi \in C^1(]0,T];C(H))$ ($\phi \in C^1([0,T];C(H))$ if $\phi_0 \in C_u^1(H)$) and

$$\frac{d}{dt}\phi(t,x) + F(\phi_x(t,x)) = 0, \quad \forall t > 0,\ x \in H \tag{3.5}$$

where $\frac{d}{dt}\phi(t,\cdot)$ is the derivative of $\phi:[0,T] \to C(H)$. We shall call such a function a *strong solution* to (3.1). Inasmuch as the operator $\phi \to -F(\phi_x)$ is dissipative in $C(H)$ (see 2.29) it follows by Lemma 1.2 in [3] (page 100) that the strong solution is unique.

We have therefore proved the following theorem.

THEOREM 1. *Let* F *satisfy assumptions* (i), (ii), (iii). *Then for each* $\phi_0 \in K$ *the Cauchy problem* (3.1) *has a unique strong solution* $\phi \in C([0,T];C(H)) \cap C(]0,T];C^1(H))$ *given by formula* (3.2). *If in addition* $\phi_0 \in C_u^1(H,\mathbb{R})$

then $\phi \in C^1([0,T];C(H)) \cap C([0,T];C^1(H))$. *Moreover the map* $\phi_0 \to \phi$ *is nonexpansive from* $C(H)$ *to* $C([0,T];C(H))$.

An interesting feature of the semigroup $S(t)$ is its *smoothing effect on initial data*, i.e., $S(t)\phi_0 \in D(L_0)$ for all $t > 0$. This property is specific to contraction semigroups generated by subgradients operators (see Proposition 9 in Chapter 1 and [3] page 188).

It is worth noticing that under assumptions of Theorem 1 we have an explicit formula for the solution to Eq. (3.1). In the special case $F(x) = \frac{1}{2}|x|^2$ the solution $\phi(t) = S(t)\phi_0$ to (3.1) is given by

$$(S(t)\phi_0)(x) = (\phi_0)_t(x) \text{ for all } t > 0 \text{ and all } x \in H \tag{3.6}$$

where $(\phi_0)_t$ is the regularized form of ϕ_0 (see Eq. (2.4) in Chapter 1). Explicit formulas for equations of this type have been given by Hopf, Lax and more recently by Conway and Hopf [13].

Now we shall study the nonhomogeneous Cauchy problem

$$\begin{aligned} &\phi_t(t,x) + F(\phi_x(t,x)) = g(t,x);\ t \in [0,T],\ x \in H \\ &\phi(0,x) = \phi_0(x) \end{aligned} \tag{3.7}$$

where F satisfies assumptions (ii), (iii).

Additionally we assume that

(v) K *is a closed convex cone of* C(H).

We shall further assume that

$$\phi_0 \in C^1(H) \cap K,\ g \in C([0,T];C^1(H) \cap K \tag{3.8}$$

where

$$K = \{\phi \in C([0,T];C(H));\ \phi(t) \in K,\ \forall t \in [0,T]\}. \tag{3.9}$$

Consider the approximating equation:

$$\begin{aligned} &\phi_t(t,x) + \varepsilon^{-1}(\phi(t,x) - S(\varepsilon)\phi(t,\cdot))(x) = g(t,x) \\ &\phi(0,x) = \phi_0(x);\ t \in [0,T],\ x \in H,\quad \varepsilon > 0 \end{aligned} \tag{3.10}$$

or equivalently

$$\phi(t,x) = \exp(-\varepsilon^{-1}t)\phi_0(x) + \int_0^t \exp(-\varepsilon^{-1}(t-s))g(s,x)ds$$
$$+ \varepsilon^{-1} \int_0^t \exp(\varepsilon^{-1}(t-s))(S(\varepsilon)\phi(s,\cdot)(x)ds. \tag{3.11}$$

By assumptions (v) and (3.8) the operator γ defined by the right-hand side of Eq. (3.11) maps K into itself and is contractant in $C([0,T];C(H))$ (with respect to every semi-norm $|\ |_{0,R}$). Thus for every $\varepsilon > 0$ Eq. (3.11) has a unique solution $\phi_\varepsilon \in K \cap C^1([0,T];C(H))$.

Since, as shown earlier, $S(\varepsilon)\phi \in C^1(H)$ for every $\phi \in K$ and $\varepsilon > 0$ we see by (3.11) that $\phi_\varepsilon \in C([0,T];C^1(H)) \cap C^1([0,T];C(H))$. Moreover, recalling that (see (2.7) and (2.20))

$$(S(\varepsilon)\phi)_x(x) = \phi'(y_\varepsilon(x)) \quad \forall x \in H,\ \varepsilon > 0 \tag{3.12}$$

we infer that $E_\varepsilon(t,x) = (\phi_\varepsilon)_x(t,x)$ is the solution to

$$\frac{d}{dt} E_\varepsilon(t,x) + \varepsilon^{-1}(E_\varepsilon(t,x) - E_\varepsilon(t,y_\varepsilon(t,x))) = g_x(t,x) \tag{3.13}$$

where

$$y_\varepsilon(t,x) + F'E_\varepsilon(t,y_\varepsilon(t,x))) = x. \tag{3.14}$$

By Eq. (3.13) it follows via Gronwall's lemma that:

$$|E_\varepsilon(t,\cdot)|_{0,r} \leq C_r(|\phi_0'|_{0,r} + |g_x(t,\cdot)|_{0,r}),\ t \in [0,T]. \tag{3.15}$$

(We shall denote by C_r several positive constants independent of ε.)

Next by (2.26), (3.12) and (3.13) we have

$$\begin{aligned} &|\varepsilon^{-1}(\phi_\varepsilon(t,x) - (S(\varepsilon)\phi_\varepsilon(t,\cdot))(x) - F(E_\varepsilon(t,x))| \\ &\leq (F'(E_\varepsilon(t,x)),E_\varepsilon(t,x) - E_\varepsilon(t,y_\varepsilon(t,x))) \\ &= -\varepsilon(F'(E_\varepsilon(t,x)), \frac{d}{dt} E_\varepsilon(t,x)) + \varepsilon(F'(E_\varepsilon(t,x)),g_x(t,x)) \\ &= \varepsilon(F'(E_\varepsilon(t,x)),g_x(t,x)) - \varepsilon\frac{d}{dt} F(E_\varepsilon(t,x)). \end{aligned} \tag{3.16}$$

Integrating on $[0,T]$ and using (3.15) we get

$$\int_0^T |\varepsilon^{-1}(\phi_\varepsilon(t,x) - (S(\varepsilon)\phi_\varepsilon(t,\cdot))(x) - F(E_\varepsilon(t,x))|dt$$

$$\leq \varepsilon\int_0^T |F'(E_\varepsilon(t,x))|\ |g_x(t,x)|\ dt \tag{3.17}$$

$$+ \varepsilon[F(\phi_0'(x)) - F(\phi_\varepsilon'(T,x))] \leq C_r\varepsilon \text{ for } x \in \Sigma_r$$

and therefore by (2.10)

$$(\phi_\varepsilon)_t(t,x) + F(\phi_\varepsilon')(t,x) = g(t,x) + \eta_\varepsilon(t,x); 0 \leq t \leq T$$

$$\phi_\varepsilon(0,x) = \phi_0(x) \tag{3.18}$$

where

$$\int_0^T |\eta_\varepsilon(t,x)\ dt \leq C_r\varepsilon, \quad \forall x \in \Sigma_r \quad \varepsilon > 0. \tag{3.19}$$

For every $g \in K$ and $\phi_0 \in K$ denote by $\phi(t,x) = G(\phi_0,g)(t,x)$ the solution to Eq. (3.7) (if any). Since the operator $\phi \to -F(\phi_x)$ is dissipative (see (2.28)) it follows that:

$$\frac{d}{dt}|G(\phi_0,g)(t,\cdot) - G(\psi_0,h)(t,\cdot)|_{0,r} \leq |g(t,\cdot) - h(t,\cdot)|_{0,r}$$

and therefore

$$|G(\phi_0,g)(t,\cdot) - G(\psi_0,h)(t,\cdot)|_{0,r} \leq |\phi_0 - \psi_0|_{0,r}$$

$$+ \int_0^t |g(s,\cdot) - h(s,\cdot)|_{0,r}\ ds \tag{3.20}$$

for all $g,h \in K$ and $\phi_0,\psi_0 \in K$. Along with (3.18) and (3.19) the latter yields:

$$|\phi_\varepsilon(t,\cdot) - \phi_\lambda(t,\cdot)|_{0,r} \leq C_r T(\varepsilon+\lambda), \text{ for all } r > 0, \varepsilon,\ \lambda > 0. \tag{3.21}$$

Hence there exists $\phi = \lim_{\varepsilon\to 0} \phi_\varepsilon$ in $C([0,T]; C(H))$. Obviously $\phi \in K$ and since by (3.13)

$$|\phi_\varepsilon'(t,\cdot)|_{0,r} \leq C_r(|\phi_0'|_{0,r} + \sup_{0\leq t\leq T} |g_x(t,\cdot)|_{0,r})$$

we conclude that for each $t \in [0,T]$, $\phi(t,\cdot) \in C_{Lip}(H)$ and for all $r > 0$

$$\|\phi(t,\cdot)\|_{0,r} \leq C_r(|\phi_0'|_{0,r} + \sup_{0 \leq t \leq T} |g_x(t,\cdot)|_{0,r}). \tag{3.22}$$

Summarizing, we have established that there exists a sequence $\{\phi_\varepsilon\} \subset C^1([0,T];C(H)) \cap C([0,T];C^1(H)) \cap K$ such that for $\varepsilon \to 0$,

$$\phi_\varepsilon \to \phi \text{ in } C([0,T];C(H)) \tag{3.23}$$

$$\{\phi_\varepsilon'\} \text{ is bounded in } C([0,T];C(H)) \tag{3.24}$$

$$(\phi_\varepsilon)_t + F(\phi_\varepsilon') \to g \text{ in } L^1(0,T;C(H)) \tag{3.25}$$

$$\phi_\varepsilon(0,x) = \phi_0(x) \text{ for all } x \text{ in } H . \tag{3.26}$$

By definition, a function ϕ satisfying the above condition is called *weak solution* to Problem (3.7). Hence for $\phi_0 \in C^1(H) \cap K$ and $g \in C([0,T];C^1(H)) \cap K$ *the Cauchy Problem (3.7) has a unique weak solution* $\phi \in K$ *and the map* $(\phi_0,g) \to \phi$ *is Lipschitz from* $C(H) \times C([0,T];C(H))$ to $C([0,T];C(H))$.

We will show now that this result is still valid if one merely assumes that $\phi_0 \in K$ and $g \in K$. To this end we fix (ϕ_0,g) in $K \times K$ and consider $(\phi_0)_\lambda$ and $\{g_\lambda\}$ the regularized of ϕ_0 and g, respectively, i.e.,

$$(\phi_0)_\lambda = S(\lambda)\phi_0, \quad g_\lambda = S(\lambda)g.$$

By Lemma 2 and the above results, $(\phi_0)_\lambda$ and g_λ satisfy conditions (3.8). Hence the problem

$$\begin{cases} \phi_t^\lambda + F(\phi_x^\lambda) = g_\lambda, & 0 \leq t \leq T,\ x \in H \\ \phi^\lambda(0,x) = (\phi_0)_\lambda(x), & x \in H \end{cases}$$

has a unique weak solution $\phi^\lambda \in K$ satisfying

$$|\phi^\lambda(t,\cdot) - \phi^\mu(t,\cdot)|_{0,r} \leq |(\phi_0)_\lambda - (\phi_0)_\mu|_{0,r}$$

$$+ \int_0^t |g_\lambda(s,\cdot) - g_\mu(s,\cdot)|_{0,r}\, ds.$$

Hence $\lim_{\lambda \to 0} \phi^\lambda = \phi$ exists in $C([0,T];C(H))$.

For any $\lambda > 0$ let $\phi^{\lambda,\varepsilon} \in C^1([0,T];C(H)) \cap K$ be the sequence satisfying conditions (3.24) $\sim$ (3.26) and such that for $\varepsilon \to 0$

$$\lim_{\varepsilon \downarrow 0} \phi^{\lambda,\varepsilon} = \phi^{\lambda} \text{ in } C([0,T];C(H)). \tag{3.27}$$

By (3.15) we see that

$$|\phi_x^{\lambda,\varepsilon}(t,\cdot)|_{0,r} \leqslant C_r(|(\phi_0)'_\lambda|_{0,r} + \sup_{0\leqslant t\leqslant T}\{|(g_\lambda)_x(t,\cdot)|_{0,r}\}$$

$$\leqslant C_r^1 \text{ for all } \lambda,\ \varepsilon > 0$$

because

$$|(\phi_0)'_\lambda|_{0,r} \leqslant \sup\{|y|;y \in \partial\phi_0(x);|x| \leqslant r\}$$

$$|(g_\lambda)_x(t)|_{0,r} \leqslant \sup\{|z(t)|;z(t) \in \partial g(t,x);|x| \leqslant r\}$$

and by Proposition 6 in Chapter 1, $\partial\phi_0$ and ∂g are bounded on bounded subsets.

We may conclude that ϕ is a weak solution to Eq. (3.7). We have therefore proved the following theorem.

THEOREM 2. *Let assumptions* (i), (ii), (iii), (v) *be satisfied. Then for all* $\phi_0 \in K$ *and* $g \in K$ *the Cauchy problem* (3.7) *has a unique weak solution* $\phi \in K$ *having the property that* $\phi(t) \in C_{Lip}(H)$ *for all* $t \in [0,T]$. *Moreover, the map* $(\phi_0,g) \to \phi$ *is Lipschitz from* $C(H) \times C([0,T];C(H))$ *to* $C([0,T];C(H))$.

REMARK 2. Assumptions (ii) and (v) are in particular satisfied by functions F of the form

$$F(x) = \zeta(|x|^2) \quad \forall x \in H \tag{3.28}$$

where ζ is a convex continuously differentiable function on $[0,\infty[$ satisfying the condition

$$\zeta'(0) > 0,\ \lim_{r\to\infty} \zeta(r)/r^{1/2} = +\infty \tag{3.29}$$

REMARK 3. By Theorem 1 it follows via Eq. (2.20) that for each $\phi_0 \in K \cap C^1(H)$

$$E(t,x) = (\nabla\phi_0^* + tF')^{-1}(x);\ t \geqslant 0,\ x \in H \tag{3.30}$$

is the solution to the *conservation law* equation

$$\begin{aligned} &E_t + (F(E))_x = 0,\ 0 < t < T,\ \ x \in H \\ &E(0,x) = \phi_0'(x). \end{aligned} \tag{3.31}$$

More will be said about this equation in §5.

§4 Existence and uniqueness for $F(x) = \frac{1}{2}|x|^2$

We shall study here the Cauchy problem:

$$\begin{aligned} &\phi_t(t,x) + \tfrac{1}{2}|\phi_x(t,x)|^2 - (Ax,\phi_x(t,x)) = g(t,x) \\ &\phi(0,x) = \phi_0(x);\ x \in D(A),\ t \in [0,T] \end{aligned} \tag{4.1}$$

where $\phi_0 \in K$ and $g \in \mathcal{K}$. Notice that in this case

$$K = \{\phi \in C(H);\ \phi \text{ convex and } 0 \in \partial\phi(0)\}$$

$$\mathcal{K} = \{\phi \in B([0,T];C(H));\ \phi(t,\cdot) \in K\}.$$

Moreover, for any $\phi \in K$ we have $S(\varepsilon)\phi = \phi_\varepsilon$, the convex regularized of ϕ (see Eq. (2.1) in Chapter 1). We list here for later use several properties of ϕ_ε.

LEMMA 3. *Assume that* $\phi,\bar{\phi} \in C^i(H) \cap K$, $i = 0,1$; *then for any* $r > 0$, $\varepsilon > 0$, *we have:*

$$|\phi_\varepsilon|_{i,r} \leqslant |\phi|_{i,r} \tag{4.2}$$

$$\|\phi_\varepsilon\|_{i,r} \leqslant \|\phi\|_{i,r} \tag{4.3}$$

$$|\phi_\varepsilon - \bar{\phi}_\varepsilon|_{i,r} \leqslant |\phi - \bar{\phi}|_{i,r} \tag{4.4}$$

Proof (4.2) and (4.3) are clear by the properties of ϕ_ε (see Chapter 1); moreover (4.4) for $i = 0$ follows from (2.16). Let us prove (4.4) for $i = 1$; we have

$$\phi_\varepsilon'(x) - \bar{\phi}_\varepsilon'(x) = \frac{1}{\varepsilon}\,[\bar{y}_\varepsilon(x) - \bar{y}_\varepsilon(x + \varepsilon(\bar{\phi}' - \phi')(y_\varepsilon(x)))]$$

recalling that $x \to \bar{y}_\varepsilon(x)$ is nonexpansive the conclusion follows.

We consider now the equation

$$\phi_t^\varepsilon(t,x) + \frac{1}{\varepsilon}(\phi^\varepsilon(t,x) - S(\varepsilon)\phi^\varepsilon(t,x)) - (Ax,\phi_x^\varepsilon(t,x))$$

$$= g(t,x);\ t \in [0,T],\ x \in D(A) \tag{4.5}$$

$$\phi^\varepsilon(0,x) = \phi_0(x).$$

We consider also a weak form of (4.5) (which is equivalent to (4.5) if A is bounded and ϕ_0, g regular),

$$\phi^\varepsilon(t,x) = e^{-t/\varepsilon}\phi_0(e^{tA}x) + \int_0^t e^{-(t-s)/\varepsilon}[\frac{1}{\varepsilon} S(\varepsilon)\phi^\varepsilon + g]$$

$$(s,\ e^{(t-s)A}x)ds; t \in [0,T],\ x \in H. \tag{4.6}$$

LEMMA 4. *Assume* (iv) *and moreover that:*

$$\phi_0 \in C^1(H) \cap K,\ g \in B([0,T];C^1(H)) \cap K \tag{4.7}$$

then the equation (4.6) *has a unique solution* $\phi^\varepsilon \in B([0,T];C^1(H)) \cap K$ *and the following estimates hold*

$$|\phi^\varepsilon(t,\cdot)|_{i,r} \leq |\phi_0|_{i,r} + \int_0^t |g(s,\cdot)|_{i,r}ds,\ i = 0,1 \tag{4.8}$$

$$|\phi^\varepsilon(t,\cdot) - \bar{\phi}^\varepsilon(t,\cdot)|_{i,r} \leq |\phi_0 - \bar{\phi}_0|_{i,r} + \int_0^t |g(s,\cdot) - \bar{g}(s,\cdot)|_{i,r}ds,\ i = 0,1 \tag{4.9}$$

(*here* $\bar{\phi}_0, \bar{g}$ *fulfil* (4.7) *and* $\bar{\phi}^\varepsilon$ *is the corresponding solution to* (4.6)).

Moreover if $x \in D(A)$, ϕ^ε *is a solution to* (4.5). *Finally if* $\phi_0 \in C^1_{Lip}(H) \cap K$ *and* $g \in B([0,T];C^1_{Lip}(H)) \cap K$ *then* $\phi^\varepsilon \in B([0,T];C^1_{Lip}(H)) \cap K$ $\forall t \in [0,T]$ *and we have:*

$$\|\phi^\varepsilon(t,\cdot)\|_{1,r} \leq \|\phi_0\|_{1,r} + \int_0^t \|g(s,\cdot)\|_{1,r}ds \tag{4.10}$$

Proof Set

$$Z = B([0,T];C^1(H)) \cap K \tag{4.11}$$

and

$$\phi^0(t,x) = e^{-t/\varepsilon}\phi_0(e^{tA}x) + \int_0^t e^{-(t-s)/\varepsilon} g(s,e^{(t-s)A}x)ds \tag{4.12}$$

$$\phi^{n+1}(t,x) = \phi^0(t,x) + \frac{1}{\varepsilon}\int_0^t e^{-(t-s)/\varepsilon}(S(\varepsilon)\phi^n(s,e^{(t-s)A}x)ds; \ n \in \mathbb{N}.$$

Then $\phi^n \in Z$ and recalling (4.4) we have

$$|\phi^0(t,\cdot)|_{i,r} \leqslant e^{-t/\varepsilon}|\phi_0|_{i,r} + \int_0^t e^{-(t-s)/\varepsilon}|g(s,\cdot)|_{i,r}ds \leqslant C_r$$

$$|\phi^{n+1}(t,\cdot) - \phi^n(t,\cdot)|_{i,r} \leqslant \frac{C_r}{\varepsilon^n}\ \frac{t^n}{n!}$$

it follows that $\{\phi^n\}$ is a Cauchy sequence in $B([0,T];C^1(H))$ which converges to the unique solution of (4.6). The inequalities (4.8) and (4.9) follow from Lemma 3 and Gronwall's lemma.

Set now $E^\varepsilon(t,x) = \phi_x^\varepsilon(t,x)$; by (2.20) we have

$$(S(\varepsilon)\phi)_x = \phi'(y_\varepsilon(x)) \quad \forall \phi \in K \cap C^1(H) \tag{4.13}$$

by (4.6) it follows

$$E^\varepsilon(t,x) = e^{-t/\varepsilon}e^{tA^*}\phi_0'(e^{tA}x) \tag{4.14}$$

$$+ \int_0^t e^{-(t-s)/\varepsilon}e^{(t-s)A^*}[\frac{1}{\varepsilon}\ E^\varepsilon(t,y_\varepsilon) + g_x](s,e^{(t-s)A}x)ds.$$

Moreover if $x \in D(A)$ we see by (4.6) that ϕ^ε is differentiable with respect to t and it is easy to check (4.5). Assume finally $\phi_0 \in C^1_{Lip}(H)$ and $g \in B([0,T];C^1_{Lip}(H))$; we have $\phi^n(t,\cdot) \in C^1_{Lip}(H)$ and by (4.3)

$$\| \phi^0(t,\cdot)\|_{1,r} \leqslant e^{-t/\varepsilon}\| \phi_0\|_{1,r} + \int_0^t e^{-(t-s)/\varepsilon}\| g(s,\cdot)\|_{1,r}ds$$

$$\|\phi^{n+1}(t,\cdot)\|_{1,r} \leqslant e^{-t/\varepsilon}\|\phi_0\|_{1,r} + \int_0^t e^{-(t,s)/\varepsilon}\|g(s,\cdot)\|_{1,r}ds$$

$$+ \frac{1}{\varepsilon}\int_0^t e^{-(t-s)/\varepsilon}\| \phi^n(s,\cdot)\|_{1,r}ds$$

from which

$$|\phi_x^n(t,x) - \phi_x^n(t,x)| \leq |x - y|\{e^{-t/\varepsilon} \sum_{k=0}^{n} \frac{t^k}{\varepsilon^k k!} \|\phi_0\|_{1,r}$$

$$+ \sum_{k=0}^{n} \int_0^t \frac{(t-s)^k}{\varepsilon^k\, k!} e^{-(t-s)/\varepsilon} \|g(s,\cdot)\|_{1,r}\, ds\}$$

and for $n \to \infty$ the conclusion follows.

Set now:

$$R_{\phi,\alpha}(x) = \frac{1}{\alpha}(\phi(x) - (S(\alpha)\phi)(x)) - \frac{1}{2}|\phi'(x)|^2 ; \tag{4.15}$$

by (2.26) we have

$$|R_{\phi,\alpha}|_{0,r} \leq \alpha\, |\phi|_{1,r}^2\, \|\phi\|_{1,r} . \tag{4.16}$$

Hence ϕ^ε satisfies the equation

$$\phi_t^\varepsilon(t,x) + \frac{1}{2}\, |\phi_x^\varepsilon(t,x)|^2 - (Ax,\phi_x^\varepsilon(t,x)) = g(t,x) + R_{\varepsilon,\phi^\varepsilon}(t,x) \tag{4.17}$$

where

$$R_{\varepsilon,\phi^\varepsilon}(t,x) = R_{\varepsilon,\phi^\varepsilon(t,\cdot)}(x);$$

and by (4.8) and (4.10), we have

$$|R_{\varepsilon,\phi^\varepsilon}(t,\cdot)|_{0,r} \leq C_{r\varepsilon}; \text{ for all } t \in [0,T] \text{ and } r > 0 . \tag{4.18}$$

We say that ϕ is a *strong solution* to (4.1) if

(a) $\phi \in B([0,T];C^1(H))$ *and* $\{\phi(t,\cdot),t \in [0,T]\}$ *is bounded in* $C^1_{Lip}(H)$

(b) $\phi \in C^1([0,T] \times D(A);\mathbb{R})$ (4.19)

(c) Eq. (4.1) *is satisfied for all* $t \in [0,T]$ *and* $x \in D(A)$.

In accordance with the definition given in §3 the function $\phi \in K$ is called *weak solution* to Eq. (4.1) if there exist the sequences $\{\phi_0^n\} \subset K$, $\{g^n\} \subset K$ and $\{\phi^n\} \subset B([0,T];C^1(H))$ such that:

(a') $\phi_0^n \to \phi_0$ *in* $C(H)$ *and* $g^n \to g$ *in* $B([0,T];C(H))$.

(b') $\phi^n \to \phi$ *in* $B([0,T];C(H))$.

(c') $\{\phi^n(t,\cdot),\ t \in [0,T],\ n \in \mathbb{N}\}$ *is bounded in* $C^1_{Lip}(H)$. (4.20)

(d') ϕ^n *are strong solutions to*

$$\phi^n_t + \frac{1}{2}|\phi^n_x|^2 - (Ax,\phi^n_x) = g^n, \phi^n(0,x) = \phi^n_0(x).$$

The main existence and uniqueness result for Eq. (4.1) is the following.

THEOREM 3. *Assume that conditions* (i), (iv) *are satisfied. Then the Cauchy Problem* (4.1) *has a unique weak solution* $\phi \in K$. *Moreover the map* $(\phi_0,g) \to \phi$ *is Lipschitz from* $C(H;\mathbb{R}) \times L^1(0,T;C(H))$ *to* $B([0,T];C(H))$.

If in addition $\phi_0 \in K \cap C^1_{Lip}(H)$ *and* $g \in K \cap B([0,T];C^1_{Lip}(H))$ *then for any* $t \in [0,T]$ *we have* $\phi(t,\cdot) \in C^1_{Lip}(H)$ *and the set* $\{\phi(t,\cdot); t \in [0,T]$ *is bounded in* $C^1_{Lip}(H)$.

Proof We begin by observing that if $\phi_{0i} \in K$, $g_i \in K$ and ϕ_i are the corresponding weak solutions to Problem (4.1) $(i = 1,2)$ then:

$$|\phi_1(t,\cdot) - \phi_2(t,\cdot)|_{0,r} \leq |\phi_{01} - \phi_{02}|_{0,r} + \int_0^t |g_1(s,\cdot) - g_2(s,\cdot)|_{0,r}ds. \tag{4.21}$$

Indeed let $\{\phi^n_i\}$ be the corresponding sequences in (4.20), then:

$$(\phi^n_i)_t + \frac{1}{\varepsilon}(\phi^n_i - S(\varepsilon)\phi^n_i) - (Ax,(\phi^n_i)_x) = g^n_i + R_{\varepsilon,\phi^n_i}$$

which is equivalent to

$$\phi^n_i(t,x) = e^{-t/\varepsilon}\phi^n_{oi}(e^{tA}x) + \int_0^t e^{-(t-s)/\varepsilon}[\frac{1}{\varepsilon} S(\varepsilon)\phi^n_i + g^n_i + R_{\varepsilon,\phi^n_i}](s,e^{(t-s)A}x)ds$$

and by the Gronwall's lemma and (4.16) we have

$$|\phi^n_1(t,\cdot) - \phi^n_2(t,\cdot)|_{0,r} \leq |\phi^n_{01} - \phi^n_{02}|_{0,r} + \int_0^t \{|g^n_1(s,\cdot) - g^n_2(s,\cdot)|_{0,r} + \varepsilon C_r\}ds$$

(4.21) follows for $n \to \infty$, $\varepsilon \to 0$ and also the last affirmation of the Theorem.

We remark now that, due to the density of $C^1_{Lip}(H) \cap K$ in K and (4.21) it suffices to prove existence for $\phi_0 \in C^1_{Lip}(H) \cap K$ and $g \in B([0,T];C^1_{Lip}(H)) \cap K$. Let ε, $\eta > 0$, we have

$$|\phi^{\varepsilon}(t,.) - \phi^{\eta}(t,.)|_{0,r} \leqslant \int_0^t e^{-(t-s)/\varepsilon}\{|(S(\varepsilon)\phi^{\varepsilon}$$

$$- S(\eta)\phi^{\varepsilon})(s,.)|_{0,r} + |(S(\varepsilon)\phi^{\varepsilon}) - (S(\eta)\phi^{\varepsilon})(s,.)|_{0,r}\}ds.$$

From (4.15), (4.16) it follows

$$|(S(\varepsilon)\phi^{\varepsilon} - S(\eta)\phi^{\varepsilon})(s,.)|_{0,r} \leqslant C_r(\varepsilon+\eta)$$

and recalling (4.4) and using the Gronwall's lemma we find

$$|\phi^{\varepsilon}(t,.) - \phi^{\eta}(t,.)|_{0,r} \leqslant C_r(\varepsilon+\eta)$$

hence $\lim_{\varepsilon\to 0} \phi_{\varepsilon} = \phi$ exists in $B([0,T];C(H))$ and ϕ is a weak solution to (4.1).

It remains to prove that $\|\phi(t,.)\|_{1,r} \leqslant C_r$. To this end we fix $r > 0$, $h \in [0,1]$ and $x,y \in \Sigma_r$. We set

$$\psi^{\varepsilon}_{x,y}(th) = \psi^{\varepsilon}(t,h) = \phi^{\varepsilon}(t,x+hy). \tag{4.22}$$

Since by Lemma 4 there exist $M_r > 0$ such that

$$|\phi^{\varepsilon}(t,\cdot)|_{0,r} + |\phi^{\varepsilon}_x(t,.)|_{0,r} + \|\phi^{\varepsilon}_x(t,\cdot)\|_{0,r} \leqslant M_r \tag{4.23}$$

we have

$$|\psi^{\varepsilon}(t,h)| \leqslant M_{2,r} \cdot \tag{4.24}$$

Also notice that

$$\frac{\partial}{\partial h}\psi^{\varepsilon}(t,h) = (y,\phi^{\varepsilon}_x(t,x+hy)) \tag{4.25}$$

and therefore

$$|\frac{\partial\psi^{\varepsilon}}{\partial h}(t,h)| \leqslant M_{2r}|y| \tag{4.26}$$

$$|\frac{\partial\psi^{\varepsilon}(t,h)}{\partial h} - \frac{\partial\psi^{\varepsilon}(t,k)}{\partial k}| \leqslant M_{2r}|h-k||y|^2. \tag{4.27}$$

Thus by the Ascoli-Arzela theorem $\{\frac{\partial\psi^\varepsilon}{\partial h}(t,h)\}$ is compact in $C([0,1])$ and for any fixed $t \in [0,T]$ we may assume that:

$$\begin{aligned} &\psi^\varepsilon(t,h) \to \phi(t,x+hy) \text{ uniformly on } [0,1] \\ &\frac{\partial\psi^\varepsilon}{\partial h}(t,h) \to \frac{d}{dh}\phi(t,x+hy) \text{ uniformly on } [0,1]. \end{aligned} \tag{4.28}$$

Hence

$$(y,\phi_x^\varepsilon(t,x+hy)) \to \frac{d}{dh}\phi(t,x+hy). \tag{4.29}$$

We may conclude therefore that $\phi(t,\cdot)$ is Gâteaux differentiable in x and:

$$\lim_{\varepsilon\to 0}(\phi_x^\varepsilon(t,x),y) = (D_x\phi(t,x),y) \tag{4.30}$$

where $D_x\phi(t,x)$ is the Gâteaux derivative of $\phi(t,\cdot)$. We will show that $D_x\phi(t,x)$ is continuous in x. Indeed we have

$$\begin{aligned} \left|\frac{\partial\psi^\varepsilon_{x,y}(0)}{\partial h} - \frac{\partial\psi^\varepsilon_{z,y}(0)}{\partial h}\right| &= (y,\phi_x^\varepsilon(t,x) - \phi_x^\varepsilon(t,z))| \\ &\leq M_r|x - z||y|. \end{aligned}$$

Hence by (4.29)

$$|(y,D_x\phi(t,x) - D_x\phi(t,z))| \leq M_r|y|\ |x - z|$$

and therefore $D_x\phi(t,\cdot) \in C^0_{Lip}(H;H)$ for all t. We have therefore proved that $D_x\phi = \phi_x$ and $\phi(t,\cdot) \in C^1_{Lip}(H)$ as claimed. This completes the proof of the Theorem.

For computational purposes, the following approximation result is useful.

PROPOSITION 1. *Let* $\{A^n\}$ *be a sequence of linear operators satisfying condition (iv) and such that* $(\lambda-A_n)^{-1} \to (\lambda-A)^{-1}$ *for all* $\lambda > 0$. *For* $\phi_0 \in C^1_{Lip}(H) \cap K$, $g \in B([0,T];C^1_{Lip}(H)) \cap K$ *let* ϕ_n *be the corresponding solution to Eq. (4.1) where* $A = A^n$. *Then* $\phi_n \to \phi$ *in* $B([0,T];C(H))$ *where* ϕ *is the solution to (4.1) corresponding to* A.

Proof Nothing is left to do except to notice that by the Trotter Theorem

$e^{tA_n}x \to e^{tA}x$ uniformly on compacts and for each $x \in H$ to use the integral representation (4.6). The details are left to the reader.

We will now find strong solutions of Problem (4.1). For this purpose we need the following lemma.

LEMMA 5. *Let* $\phi \in C^2_{Lip}(H) \cap K$; *then for any* $r > 0$, $\varepsilon > 0$

$$|\phi_\varepsilon|_{2,r} \leq |\phi|_{2,r};\ \|\phi_\varepsilon\|_{2,r} \leq \|\phi\|_{2,r} \tag{4.31}$$

$$R_{\phi,\varepsilon}|_{1,r} \leq \varepsilon\{\tfrac{1}{2}|\phi|^2_{1,r}\ \|\phi\|_{2,r} + |\phi|^2_{2,r}|\phi|_{1,r}\} \tag{4.32}$$

Proof We set $E = \phi_x$, then $y'_\varepsilon(x) = (1+\varepsilon E'(y_\varepsilon(x)))^{-1}$; $E'(y_\varepsilon(x))$ being a positive operator (4.31) follows. Moreover (4.32) follows by

$$R'_{\phi,\varepsilon}(x) = \frac{1}{\varepsilon}(E(x) - E(y_\varepsilon(x))) - E'(x)\cdot E(x)$$

$$= \int_0^1 E'(\zeta x + (1-\zeta)y_\varepsilon(x))\cdot E(y_\varepsilon(x))d\zeta - E'(x)\cdot E(x).$$

THEOREM 4. *Assume that conditions (i), (iv) are satisfied and that* $\phi_0 \in K \cap C^2_{Lip}(H)$, $g \in K \cap B([0,T];C^2_{Lip}(H))$. *Then Problem (4.1) has a unique strong solution* $\phi \in B([0,T];C^1(H))$; *moreover for any* $t \in [0,T]$, $\phi(t,\cdot) \in C^2_{Lip}(H)$ *and* $\{\phi(t,\cdot);\ t \in [0,T]\}$ *is bounded in* $C^2_{Lip}(H)$.

Proof Let $\phi^{n,\varepsilon}$ defined as (4.12); then by Lemma 5 we get the estimation:

$$\|\phi^{n,\varepsilon}(t,\cdot)\|_{2,r} \leq \|\phi_0\|_{2,r} + \int_0^t \|g(s,\cdot)\|_{2,r}\ ds. \tag{4.33}$$

Now, arguing as in the last part of the proof of Theorem 3 we see that $\|\phi(t,\cdot)\|_{2,r} \leq C_r$ and by (4.32) we find

$$|R_{\varepsilon,\phi_\varepsilon}|_{1,r} \leq \varepsilon C_r. \tag{4.34}$$

Let now ε, $\eta > 0$, by (4.32) and (4.33) we get

$$|\phi^\varepsilon(t,\cdot) - \phi^\eta(t,\cdot)|_{1,r} \leq (\varepsilon+\eta)C_r \tag{4.35}$$

hence $\lim_{\varepsilon\to 0} \phi^\varepsilon = \phi$ exists in $B([0,T];C^1(H))$ and the conclusion follows from (4.17).

We will study now Eq. (1.6); and for this we need some notations. Let Π be the convex cone of all $E \in C(H;H)$ which are monotone on H and satisfy the condition $E(0) = 0$. We denote by $E = E(1 + \varepsilon E)^{-1}$ the Yosida approximation of E and notice that

$$E'_\varepsilon(x) = E'((1 + \varepsilon E)^{-1}x)(1 + \varepsilon E'((1 + \varepsilon E)^{-1}x))^{-1}. \tag{4.36}$$

Then by an easy calculation which uses the fact that the operator $(1 + \varepsilon E)^{-1}$ is nonexpansive in H we may prove:

LEMMA 6 *Let* $E \in \Pi$. *Then for all* $\varepsilon > 0$ *and* $r > 0$ *one has:*

$$|E_\varepsilon|_{0,r} \leq |E|_{0,r} \tag{4.37}$$

$$|E_\varepsilon|_{1,r} \leq |E|_{1,r} \text{ if } E \in C^1(H;H) \tag{4.38}$$

$$\|E_\varepsilon\|_{1,r} \leq \|E\|_{1,r} + \varepsilon|E|_{1,r} \text{ if } E \in C^1_{Lip}(H;H) \tag{4.39}$$

Moreover if $|E_i|_{0,r} + \|E_i\|_{o,r} + \|E_i\|_{1,r} \leq \alpha$, $i = 1,2$, *then there exists* $\eta(\alpha) > 0$ *such that:*

$$|(E_1)_\varepsilon - (E_2)_\varepsilon|_{1,r} \leq (1 + \varepsilon\eta(\alpha))|E_1 - E_2|_{1,r} \tag{4.40}$$

$$\left|\frac{E-E_\varepsilon}{\varepsilon} - E_x\, E\right|_{0,r} \leq \eta(\alpha)\varepsilon. \tag{4.41}$$

We consider the approximating equation

$$\begin{aligned} E^\varepsilon(t,x) &= e^{-t/\varepsilon}e^{tA^*}E_0(e^{tA}x) \\ &+ \int_0^t e^{(t-s)/\varepsilon}e^{(t-s)A^*}[\tfrac{1}{\varepsilon}\, E^\varepsilon_\varepsilon + G](s,e^{(t-s)A}x)ds. \end{aligned} \tag{4.42}$$

LEMMA 7. *Assume (iv) and that*

$$E_0 \in C^1_{Lip}(H;H) \cap \Pi,\ G_o \in B([0,T];C^1_{Lip}(H;H)), \tag{4.43}$$

$$G(t,.) \in \Pi \quad \forall t \in [0,T]$$

Then Equation (4.42) has a unique solution $E^{\varepsilon} \in B([0,T];\ C^1(H;H))$ *with* $E^{\varepsilon}(t,\cdot) \in \Pi \quad \forall t \in [0,T]$. *Moreover, the following estimates hold*

$$|E^{\varepsilon}(t,\cdot)|_{i,r} \leq |E_0|_{i,r} + \int_0^t |G(s,\cdot)|_{i,r} ds,\ i = 0,1 \tag{4.44}$$

$$\|E^{\varepsilon}(t,\cdot)\|_{1,r} \leq e^{tC_r} \|E_0\|_{1,r} + \int_0^t e^{(t-s)C_r} \|G(s,\cdot)\|_{1,r} ds \tag{4.45}$$

where C_r *is a suitable constant.*

Proof Set

$$E^0(t,x) = e^{-t/\varepsilon} e^{tA^*} E_0(e^{tA}x)$$
$$+ \int_0^t e^{-(t-s)/\varepsilon} e^{(t-s)A^*} G(s,e^{(t-s)A}x)ds \tag{4.46}$$

$$E^{n+1}(t,x) = E^0(t,x) + \frac{1}{\varepsilon} \int_0^t e^{-(t-s)/\varepsilon} e^{(t-s)A^*} E^n_{\varepsilon}(s,e^{(t-s)A}x)ds.$$

By Lemma 6 we deduce that there exists a positive constant $\beta(r,E_0,G) = \beta(r)$ such that

$$|E^n(t,\cdot)|_{0,r} + |E^n(t,\cdot)|_{1,r} + \|E^n(t,\cdot)\|_{1,r} \leq \beta(r).$$

Moreover by (4.40) we get

$$|E^{n+1}(t,\cdot) - E^n(t,\cdot)|_{1,r} \leq \frac{1}{\varepsilon} \int_0^t e^{-(t-s)/\varepsilon}(1+\varepsilon\eta(\beta(r)))\cdot$$
$$\cdot\ |E^n(s,\cdot) - E^{n-1}(s,\cdot)|_{1,r} ds. \tag{4.47}$$

It follows that $\{E^n\}$ is a Cauchy sequence in $B([0,T];C^1(H;H))$ which converges to the unique solution to Eq. (4.42). The inequalities (4.46) and (4.47) follow via Gronwall's lemma and Lemma 6.

We remark now that E^{ε} satisfies the equation

$$E^{\varepsilon}(t,x) = e^{-t/\varepsilon} e^{tA^*} E_0(e^{tA}x)$$
$$+ \int_0^t e^{(t-s)/\varepsilon} e^{(t-s)A^*} [E^{\varepsilon} + G + R'_{\varepsilon,E^{\varepsilon}_{\varepsilon}}](s,e^{(t-s)A}x)ds$$

where, due to (4.41), we have:

$$|R'_{\varepsilon,E_\varepsilon^\varepsilon}(t,\cdot)|_{0,r} \leq \varepsilon C_r. \tag{4.48}$$

We say that E is a *mild solution* to Eq. (1.6) if

(a) $E \in B([0,T];C^1(H;H))$ *and* $\{E(t,\cdot),\ t \in [0,T]\}$ *is bounded in* $C^1_{Lip}(H;H)$.

(b) *We have*

$$E(t,x) = e^{-t/\varepsilon} e^{tA^*} E_0(e^{tA}x) + \int_0^t e^{-(t-s)/\varepsilon} e^{(t-s)A^*} \tag{4.49}$$

$$[E_x \cdot E + G](s,e^{(t-s)A}x)ds.$$

Moreover, the function E is called a *weak solution* to Eq. (1.6) if there exist the sequences $\{E_0^n\} \subset \Pi$, $\{G^n\} \subset B([0,T];C(H;H))$ with $G^n(t,\cdot) \in \Pi$ such that

(a') $E_0^n \to E_0$ *in* $C(H;H)$ *and* $G^n \to G$ *in* $B([0,T];C(H;H))$.

(b') $G^n \to G$ *in* $B([0,T];C(H;H))$. (4.50)

(c') $\{E^n(t,\cdot),\ t \in [0,T],\ n \in \mathbb{N}\}$ *is bounded in* $C^1_{Lip}(H;H)$.

(d') E^n *are mild solution to* Eq. (1.6) *with* $G = G_n$ *and* $E_0 = E_0^n$.

We can prove now:

THEOREM 5. *Assume that conditions* (i), (iv) *are satisfied. Then the Cauchy Problem (1.6) has at most one weak solution. If in addition* $E_0 \in \Pi \cap C^1_{Lip}(H;H)$, $G \in B([0,T];C^1_{Lip}(H;H))$ *and* $G(t,\cdot) \in \Pi\ \ \forall t \in [0,T]$, *Problem (1.6) has a weak solution* E. *Moreover,* $\{E(t,\cdot),\ t \in [0,T]\}$ *is bounded in* $C^1_{Lip}(H;H)$. *Finally, the map* $(E_0,G) \to E$ *is Lipschitz from* $C(H;H) \times L^1(0,T;C(H;H))$ *to* $B([0,T];C(H;H))$.

Proof We have only to repeat the proof of Theorem 3 using Lemmas 6 and 7.

As regards the mild solution to Eq. (1.6) we notice that by the previous methods we can prove the following regularity result:

THEOREM 6. *Assume that conditions (i), (iv) are satisfied and that* $E_0 \in C^2_{Lip}(H;H) \cap \Pi$, $G \in B([0,T];C^2_{Lip}(H;H))$ *with* $G(t,\cdot) \in \Pi\ \forall t \in [0,T]$. *Then Eq. (1.6) has a unique mild solution* $E \in B([0,T];C^1(H:H))$, *moreover* $\{E(t,\cdot),\ t \in [0,T]\}$ *is bounded in* $C^2_{Lip}(H;H)$.

The proof is identical with that of Theorem 5 and so will be omitted.

REMARK 5. Assume, instead of (iv),

$$|e^{tA}| \leqslant e^{\omega t}, \quad \omega > 0 \tag{4.51}$$

if we set $\chi(t,x) = \phi(t,e^{-\omega t}x)$ then Eq. (4.1) reduces to

$$\chi_t(t,x) + b(t)|\chi_x(t,x)|^2 - (A_1x,\chi_x(t,x)) = g_1(t,x) \tag{4.52}$$

where $b(t) = e^{2\omega t}$, $A = A - \omega$, $g_1(t,x) = g(t,e^{-\omega t}x)$.

The approximating equation is

$$\chi_t^\varepsilon + \frac{b(t)}{\varepsilon}(\chi^\varepsilon - S(\varepsilon)\chi^\varepsilon) - (A_1x,\chi_x^\varepsilon) = g_1 \tag{4.53}$$

which we can write in weak form

$$\begin{aligned}\chi^\varepsilon(t,x) &= e^{-B(t)/\varepsilon}\phi_0(e^{tA}x)\\ &+ \int_0^t e^{-(B(t)-B(s))/\varepsilon}[\frac{1}{\varepsilon}\, b(s)\ \chi_\varepsilon^\varepsilon + g_1](s,e^{(t-s)A}x)ds\end{aligned} \tag{4.54}$$

where $B(t) = \int_0^t b(s)ds$.

Now we can repeat the previous arguments.

§5 Further regularity results

We shall give here some conditions in order to show the solution ϕ to Eq. (4.1) belongs to $C^1([0,T] \times H)$. We assume here:

$$\begin{aligned}&\text{(i)}\quad \phi_0 \in K \cap C^2_{Lip}(H).\\ &\text{(ii)}\quad \forall x \in H,\ \phi_0(x) \in D(A^*) \text{ and } A^*\phi_0 \in C^2_{Lip}(H).\\ &\text{(iii)}\quad g \in B([0,T];C^2_{Lip}(H)),\ g(t,.) \in K\ \ \forall t \in [0,T]\\ &\qquad \text{and } g_t \in B([0,T];C^1(H)).\end{aligned} \tag{5.1}$$

By virtue of Theorem 4 there exists the unique strong solution ϕ of Problem (4.1).

If we set formally $F = \phi_t$, then we find:

$$\begin{aligned}&F_t + (\phi_x - A_x,F_x) = g_t\\ &F(0,x) = F_0(x)\end{aligned} \tag{5.2}$$

where:

$$F_0(x) = -\frac{1}{2}|\phi_0'(x)|^2 + (x, A^*\phi_0'(x)) + g(0,x) \tag{5.3}$$

due to (5.1) we have $F_0 \in C^1(H)$.

The characteristics equation associated to Problem (5.2) is

$$\begin{cases} \zeta' - \phi_x(t,\zeta) + A\zeta = 0 \\ \zeta(s) = x \end{cases} \tag{5.4}$$

which has a unique global solution $\zeta(s,t,x)$ for each $x \in H$ and $0 \leqslant t \leqslant s \leqslant T$. Thus we have the following lemma:

LEMMA 8. *Under hypotheses (5.1) Problem (5.2) has a unique solution* $F \in C^1([0,T] \times H)$, *given by*

$$F(t,x) = F_0(\zeta(0,t,x)) + \int_0^t g_t(s,\zeta(s,t,x))ds. \tag{5.5}$$

Set now $F^\varepsilon(t,x) = \phi_t^\varepsilon(t,x)$; to compute $\phi_{tt}^\varepsilon = F_t^\varepsilon$ we need the following lemma:

LEMMA 9. *If* $\phi \in C^1([0,T] \times H) \cap K$ *we have*

$$\frac{d}{dt}(\phi_\varepsilon(t,x)) = \phi_t(t,y_\varepsilon(t,x)) \tag{5.6}$$

Proof By (2.13) we have

$$\phi_\varepsilon(t,x) = \phi(y_\varepsilon(t,x)) + \frac{t}{2}|\phi'(t,y_\varepsilon(t,x))|^2$$

from which (5.6) follows, recalling the equality

$$\frac{dy_\varepsilon(t,x)}{dt} + \alpha\phi'(t,y_\varepsilon(t,x)) = \frac{dy_\varepsilon(t,x)}{dt} + tE_t(t,y_\varepsilon(t,x))$$

where $E = \phi_x$.

Now taking the derivative in t in Eq. (4.5), we have

$$F_t^\varepsilon(t,x) + \frac{1}{\varepsilon}(F^\varepsilon(t,x) - F^\varepsilon(t,y_\varepsilon(t,x))) - (Ax, F_x^\varepsilon(t,x)) = g_t(t,x)$$

$$F^\varepsilon(0,x) = F_0^\varepsilon(x) = \frac{1}{\varepsilon}(\phi_0(x) - (S(\varepsilon)\phi_0)(x)) + \qquad (5.7)$$

$$- (x, A^*\phi_0'(x)) + g(0,x).$$

THEOREM 7. *Assume that (i), (iv) and (5.1) hold. Let* ϕ *be the strong solution to Problem (4.1). Then* $\phi \in C^1([0,T] \times H)$.

Proof We can easily solve linear Problem (5.7) and prove that $F^\varepsilon \to F$ in $B([0,T];C^1(H;\mathbb{R}))$; it follows:

$$\begin{cases} F_t^\varepsilon = \phi_t^\varepsilon \xrightarrow{\varepsilon>0} F \\ \phi^\varepsilon \to \phi \end{cases} \quad \text{in } B([0,T];C^1(H;\mathbb{R})) \quad .$$

This implies that $\phi \in C^1([0,T] \times H)$.

We study now regularity of solutions to Eq. (1.6); for this purpose we consider the semigroup in $C(H;H)$ defined by

$$G_t(E)(x) = e^{tA^*}E(e^{tA}x), \quad \forall E \in C(H;H). \qquad (5.8)$$

Notice that G_t maps the cone Π into itself. Define the operators

$$\mathbb{A}(E)\cdot x = \lim_{t\downarrow 0} \frac{1}{t}(G_t(E)\cdot x - Ex); \; E \in D(\mathbb{A})$$

where

$$D(\mathbb{A}) = \{E \in C(H,H); \; \exists \lim_{t\downarrow 0} \frac{1}{t}(G_t(E)x - Ex) \in C(H,H)\}$$

LEMMA 10. *Let* $E \in D(\mathbb{A}) \cap C^1(H,H)$. *Then for each* $x \in D(A)$ $E(x) \in D(A^*)$ *and*

$$\mathbb{A}(E)x = \mathbb{A}^*E(x) + E'(x)Ax \qquad (5.9)$$

Proof We set $g = \mathbb{A}(E)$ and take $x,y \in D(A)$. We have

$$(g(x),y) = \frac{d}{dh}(e^{hA^*}E(e^{hA}x),y)_{h=0}$$

$$= \frac{d}{dh} (E(e^{hA}x), E(e^{hA}y))_{h=0}$$

$$= (E'(x)Ax, y) + (E(x), y).$$

Hence the linear operator $y \to (E(x), Ay) = (g(x), y) + (E'(x)Ax, y)$ is continuous on H. In other words, $E(x) \in D(A^*)$ and (5.9) holds.

Now we may write Eq. (1.6) in terms of $\mathbb{A}$, as

$$E_t - \mathbb{A}(E) + E_x \cdot E = G$$

$$E(0) = E_0 \tag{5.10}$$

A function $E \in B([0,T];C^1(H,H)) \cap B([0,T]; D(\mathbb{A}))$ which satisfies (5.10) on $[0,T]$ is called *classical solution* to Eq. (1.6).

THEOREM 8. *Assume (i), (iv) and in addition*

$$E_0 \in C^2_{Lip}(H;H) \cap D(\mathbb{A}) \cap \Pi$$

$$\mathbb{A}(E_0) \in C^1(H;H)$$

$$G, G_t \in B([0,T];C^2_{Lip}(H;H))$$

Then Problem (1.6) has a unique classical solution.

Proof Let $V \in B([0,T];C^1(H;H))$ the 'mild' solution to the linear equation

$$V_t - \mathbb{A}(V) + V_x E + E_x V = G$$

$$V(0) = G(0) + \mathbb{A}(E_0) - E_0' \cdot E_0$$

where $E \in B([0,T];C^1(H;H))$ is the 'mild' solution to Eq. (1.6). We set $\mathbb{A}_n(E) = A_n^* E + E_x A_n$ where A_n is the Yosida approximating of A, and consider the system

$$E_t^n - \mathbb{A}_n(E^n) + E_x^n \cdot E^n = G$$

$$E^n(0) = E_0$$

$$V_t^n - \mathbb{A}_n(V^n) + V_x^n V^n + E_x^n \cdot V^n = G_t$$

$$V^n(0) = G(0) + \mathbb{A}_n(E_0) - E_0' \cdot E_0$$

which obviously has a unique classical solution (E^n, V^n). It is also clear that $V^n = E^n_t$ and by a standard procedure we can prove that

$$V^n \to V,\ E^n \to E \text{ in } B([0,T];C^1(H;H)).$$

Hence $V = E_t$ and the claim follows.

§6 An application to construction of optimal feedback controllers

Consider the optimal control problem: minimize

$$\int_0^T (g(x(t)) + \frac{1}{2}|u(t)|^2)dt + \phi_0(x(T)) \tag{6.1}$$

over all $u \in L^2(0,T;H)$ and $x \in C([0,T];H)$ subject to:

$$\begin{aligned} x' &= Ax + u \\ & \qquad\qquad t \in [0,T] \\ x(0) &= x_0. \end{aligned} \tag{6.2}$$

Here A is the infinitesimal generator of a C_0-semigroup of linear contractions on H and $g:H \to \mathbb{R}^+$, $\phi_0:H \to \mathbb{R}^+$ are convex continuous functions satisfying the condition

$$\phi_0,\ g \in K \cap C^1_{Lip}(H). \tag{6.3}$$

By the solution to (6.2) we shall mean of course the 'mild' solution (see Eq. (4.1)' in Chapter 1).

We associate the following equation (the Bellaman equation of Problem (6.1)) with Problem (6.1):

$$\begin{aligned} &\psi_t(t,x) - \frac{1}{2}|\psi_x(t,x)|^2 + (Ax,\psi_x(t,x)) + g(x) = 0; \\ &\qquad\qquad\qquad\qquad t \in [0,T],\ x \in D(A) \\ &\psi(T,x) = \phi_0(x). \end{aligned} \tag{6.4}$$

It must be observed that by the substitution $\phi(t,x) = \psi(T-t,x)$, Eq. (6.4) reduces to Eq. (4.1).

By virtue of Theorem 3, Problem (6.4) has a unique weak solution ψ; moreover there exists a sequence $\{\psi^\varepsilon\} \subset B([0,T];C^1(H))$ such that:

(a) $\psi^\varepsilon \to \psi$ in $B([0,T];C(H))$

(b) $\| \psi^\varepsilon(t,\cdot)\|_{1,r} \leq C_r$.

(c) $$\psi_t^\varepsilon - \frac{1}{2}|\psi_x^\varepsilon|^2 + (A_x,\psi_x^\varepsilon) + g = \eta^\varepsilon,$$
$$\psi^\varepsilon(T,x) = \phi_0(x), \tag{6.5}$$
$$|\eta^\varepsilon(t,\cdot)|_{0,r} \leq C_r'\varepsilon$$

(d) $(z,\psi_x^\varepsilon(t,x)) \to (z,\psi_x(t,x))$ for all $z \in H$.

It turns out that this solution is just the optimal value function of Problem (6.1). More precisely we have:

PROPOSITION 2. *Let* ψ *be the weak solution to Cauchy Problem (6.4). Then for all* $t \in [0,T]$ *and* $x_0 \in H$

$$\psi(t,x_0) = \inf \{\int_t^T (g(x(s)) + \frac{1}{2}|u(s)|^2) + \phi_0(x(T)); \tag{6.6}$$
$$x' = Ax + u,\ t \leq s \leq T,\ x(t) = x_0,\ u \in L^2(t,T;H)\}.$$

Furthermore

$$u(t) = -\psi_x(t,x(t)) \qquad 0 \leq t \leq T \tag{6.7}$$

is the optimal feedback control (unique) of Problem (6.1).

Proof Let $t \in [0,T]$ and $u \in L^2(t,T;H)$ be arbitrary but fixed. Let $x \in C([t,T];H)$ be the solution ('mild') to (6.2) on $[t,T]$, i.e.,

$$x(s) = e^{(s-t)A}x_0 + \int_t^s e^{(\tau-t)A}u(\tau)d\tau, \quad t \leq s \leq T.$$

Set $x_{0n} = n(nI-A)^{-1}x_0$, $u_n = n(nI-A)^{-1}u$ and let x_n be the corresponding solution to (6.2); we have $x_n \in C^1([t,T];H) \cap C([t,T];D(A))$ and $x_n \to x$ in $C([t,T];H)$. On the other hand, the obvious equality

$$\frac{d}{ds}\psi^\varepsilon(s,x_n(s)) = \psi^\varepsilon(s,x_n(s)) + (x_n'(s), \psi_x^\varepsilon(s,x_n(s))$$

yields

$$\frac{d}{ds}\,\psi^{\varepsilon}(s,x_n(s)) = \frac{1}{2}\,|\psi_x^{\varepsilon}(s,x_n(s)) + u_n(s)|^2 - g(x_n(s)) +$$

$$- \frac{1}{2}\,|u_n(s)|^2 + \eta^{\varepsilon}(s,x_n(s)).$$

Integrating in [t,T] and letting n tend to infinity we have

$$\psi^{\varepsilon}(t,x_0) + \frac{1}{2}\int_t^T |\psi_x^{\varepsilon}(s,x(s)) + u(s)|^2\, ds$$
$$= \int_t^T [g(x(s)) + \frac{1}{2}\,|u(s)|^2]ds + \phi_0(x(T)) - \int_t^T \eta^{\varepsilon}(s,x(s))ds. \tag{6.8}$$

Letting ε tend to zero we get

$$\psi(t,x_0) \leqslant \int_t^T (g(x(s)) + \frac{1}{2}\,|u(s)|^2)\, ds + \phi_0(x(T))$$

and therefore

$$\psi(t,x_0) \leqslant \inf \{\int_t^T (g(x(s)) + \frac{1}{2}\,|u(s)|^2)ds + \phi_0(x(T));$$
$$x' = Ax + u,\ t \leqslant s \leqslant T,\ \ x(t) = x_0,\ u \in L^2(t,T;H)\} = J(t,x_0). \tag{6.9}$$

Now consider the Cauchy problem

$$x_{\varepsilon}' - Ax_{\varepsilon} = \psi_x^{\varepsilon}(s,x)$$
$$x_{\varepsilon}(t) = x_0. \tag{6.10}$$

For each $\varepsilon > 0$, Problem (6.10), has a unique solution $x^{\varepsilon} \in C([0,T];H)$. Because $|\psi_x^{\varepsilon}(t,\cdot)|_{0,r} \leqslant C_r$ and $(z,\psi_x^{\varepsilon}(t,z)) \geqslant 0\ \forall z \in H$ it follows by a standard device that the solution x^{ε} is global. Set in (6.8) $u_{\varepsilon} = -\psi_x^{\varepsilon}(s,x_{\varepsilon})$; then

$$\psi^{\varepsilon}(t,x_0) = \int_t^T \{g(x_{\varepsilon}(s)) + \frac{1}{2}\,|u_{\varepsilon}(s)|^2\}\, ds + \phi_0(x_{\varepsilon}(T)) +$$
$$- \int_t^T \eta_{\varepsilon}(s,x_{\varepsilon}(s))ds \tag{6.11}$$

from which

$$\psi^{\varepsilon}(t,x_0) + \int_t^T \eta_{\varepsilon}(s,x_{\varepsilon}(s))ds \geqslant J(t,x_0) \tag{6.12}$$

for $\varepsilon \to 0$ we get $\psi(t,x_0) \geqslant J(t,x_0)$ because $|x_{\varepsilon}(t)| \leqslant |x_0|$ and $|\eta_{\varepsilon}(s,x_{\varepsilon}(s))| \leqslant \varepsilon C'_{|x_0|}$. So we have proved that $\psi(t,x_0)$ is the value function.

Let (x^*,u^*) be the optimal pair of Problem (6.1) (obviously such an optimal pair exists and it is unique). Setting in (6.8) $u = u^*$, $x = x^*$ and letting ε tend to zero we have

$$\lim_{\varepsilon\to 0} \int_t^T \quad |\psi_x^{\varepsilon}(s,x^*(s)) + u^*(s)|^2 ds = 0 \tag{6.13}$$

that is $\psi_x^{\varepsilon}(\cdot,x^*) \to - u^*$ in $L^2(t,T;H)$; recalling that ψ_x^{ε} converges weakly to ψ_x we get finally $u^*(s) = - \psi_x(s,y^*(s))$.

REMARK 6. By Proposition 2 it follows that the weak solution $\phi:[0,\infty[\times H \to \mathbb{R}$ to the Cauchy problem:

$$\begin{aligned} &\phi_t + \frac{1}{2} |\phi_x|^2 - (Ax,\phi_x) = g \text{ for } t \geqslant 0, \quad x \in H \\ &\phi(0,x) = \phi_0(x) \end{aligned} \tag{6.14}$$

is expressed as

$$\begin{aligned} &\phi(t,x_0) = \inf \{\int_0^t (g(x(s)) + \frac{1}{2} |u(s)|^2 ds + \phi_0(x(t)); \\ &x' = Ax + u, \ x(0) = x_0, \text{ for } t \geqslant 0, \ x_0 \in H. \end{aligned} \tag{6.15}$$

REMARK 7. (a) The optimal control u^* is given by $- \psi_x(t,x^*)$ where x^* is the solution to the closed loop differential equation

$$\begin{aligned} &x'(t) = Ax(t) - \psi_x(t,x(t)) \\ &x(0) = x_0. \end{aligned} \tag{6.16}$$

If the space H is separable we can solve (6.16) by the usual methods; we have only to remark that the mapping $s \to \psi_x(s,y(s))$ is measurable for each $y \in C([0,T],H)$ because $(\psi_x(s,y(s),z) = \lim_{\varepsilon\to 0} (\psi_x^{\varepsilon}(s,y(s)),z)$ for each $z \in H$.

(b) If we assume $\phi_0 \in C(H) \cap K$ and $g \in B([0,T]), C(H)) \cap K$ (instead of hypothesis (6.3)) then the previous arguments work (which are straightforward modifications). In this case the optimal control u^* is related to the optimal state y^* by the formula

$$u^*(t) \in -\partial\psi(t,y^*(t))$$

where $\partial\psi$ is the subdifferential of $\psi(t,\cdot)$.

§7 A variational approach to the Hamilton-Jacobi equations

We shall study here Eq. (1.1) for $F:H \to \mathbb{R}$ of the form

$$F(x) = \Phi(Dx), \quad x \in H \tag{7.1}$$

where $D \in \mathcal{L}(H;U)$ and $\Phi:U \to \mathbb{R}$. Here U is another real Hilbert space with the norm denoted $\| \ \|$ and inner product $\langle\cdot,\cdot\rangle$. More precisely the following conditions will be imposed

(a) $\Phi \in C^1(U;\mathbb{R})$ *is a convex function on* U *satisfying*

$$\Phi(0) = 0; \ \Phi(u) \leq C(\|u\|^2 + 1) \text{ for all } u \in U \tag{7.2}$$

(b) $g:H \to \mathbb{R}^+$ *and* $\phi_0:H \to \mathbb{R}^+$ *are convex continuous functions which are bounded on every bounded subset.*

(c) A *is the generator of a* C_0*-semigroup* e^{tA} *on* H.

The main result is

THEOREM 9. *Under assumptions (a), (b), (c) there exists at least one function* $\phi:[0,T] \times H \to \mathbb{R}$ *such that*

$$\phi \in C([0,T] \times H) \text{ and } \|\phi(t,\cdot)\|_{0,r} \text{ is bounded for any } r. \tag{7.3}$$

$$\text{For every } t \in [0,T], \ \phi(t,\cdot) \text{ is a convex function} \tag{7.4}$$

$$\text{For every } x \in D(A), \ \phi(\cdot,x) \text{ is Lipschitz on } [0,T] \tag{7.5}$$

$$\phi_t(t,x) + \Phi(D\partial\phi(t,x)) - (Ax,\partial\phi(t,x)) = g(x) \tag{7.6}$$

a.e. $t \in]0,T[, \ x \in D(A)$

$$\phi(0,x) = \phi_0(x) \text{ for all } x \in H. \tag{7.7}$$

In Eq. (7.6) we have denoted by ϕ_t the derivative of ϕ as function of t (which exists a.e. on]0,T[) and by $\partial\phi$ the subdifferential of $\phi(t,\cdot)$. As a matter of fact the exact meaning of Eq. (7.6) is that there exists $\zeta(t,x) \in \partial\phi(t,x)$ such that

$$\phi_t(t,x) + \Phi(D\zeta(t,x)) - (Ax,\zeta(t,x)) = g(x) \tag{7.8}$$

for all $x \in D(A)$ and a.e. $t \in [0,T]$

To prove the theorem we define the function $\psi:[0,T] \times H \to \mathbb{R}$,

$$\psi(t,x_0) = \inf \{\int_t^T (g(x(s)) + h(u(s)))ds + \phi_0(x(T));$$

$$u \in L^2(t,T;U);\ x' = Ax - D^*u,\ x(t) = x_0\} \tag{7.9}$$

where $h = \Phi^*$ (the conjugate function of Φ) and D^* is the adjoint of D. Here and throughout the following text by solution to $x' = Ax - D^*u$ we shall mean the 'mild' solution). We notice that by (7.2),

$$h(u) \geq \gamma \|u\|^2 + C \ \forall u \in U \tag{7.10}$$

for some $\gamma > 0$ and $C \in \mathbb{R}$.

LEMMA 11. *For all* $(t,x_0) \in [0,T] \times H$ *the infimum defining* $\psi(t,x_0)$ *is attained in at least one pair* (x^*,u^*). *Moreover for every* $t \in [0,T]$ *the function* $\psi(t,\cdot)$ *is convex and locally Lipschitz on* H. *For every* $x_0 \in D(A)$ *the function* $\psi(\cdot,x_0)$ *is Lipschitz on* [0,T].

Proof Let us denote by $x(s) = x(s,t,x_0,u)$ the solution to

$$\begin{aligned} &x' = Ax - D^*u \quad \text{a.e. on }]t,T[\\ &x(t) = x_0 \end{aligned} \tag{7.11}$$

By (7.10) we see that for $\|u\|_{L^2(t,T;U)} \to +\infty$,

$$\psi^t = \int_t^T (g(x(s,t,x_0,u)) + h(u(s)))ds + \phi_0(x(T,t,x_0,u)) \to +\infty.$$

Hence the convex continuous function $\psi^t:L^2(t,T;U) \to \mathbb{R}$ attains its infimum in at least one point u^t. Since g,h and ϕ_0 are convex we may infer that for all $t \in [0,T]$, $\psi(t,\cdot)$ is convex on H. If (x^*,u^*) is an optimal pair in Problem (7.9) (i.e. $x^*(s) = x(s,t,x_0,u^*)$) then by Proposition 10 in Chapter 1 there exists $p^t \in C([t,T];H)$ such that

$$(p^t)' + A^*p^t \in \partial g(x^t) \quad s \in [t,T] \tag{7.12}$$

$$p^t(T) + \partial\phi_0(x^*(T)) \ni 0, \tag{7.13}$$

$$Dp^t(s) + \partial h(u^*(s)) \ni 0, \text{ a.e. } s \in]t,T[\ . \tag{7.14}$$

It is useful to notice that since for $|x_0| \leq r$, $\psi(t,x_0) \leq M_r$ where M_r is independent of t, by (7.10) it follows that there exists C independent of t such that for all r and $|x_0| \leq r$

$$|x^t(s)|^2 + \int_t^T \|u^t(s)\|^2 ds \leq C_r, \text{ for all } s \in [t,T]. \tag{7.15}$$

Let $0 \leq t \leq t_1 \leq T$ and let $v:[t,T] \to U$ be defined by (u_0 is an element of U such that $h(u_0) < \infty$)

$$v(s) = u_0 \text{ for } t \leq s \leq t_1; \ v = u^{t_1} \text{ on } [t_1,T].$$

Then we have

$$\begin{aligned} 0 \leq \psi(t,x_0) - \psi(t_1,x_0) &\leq \int_t^{t_1} (g(x(s,t,x_0,u_0)) \\ &+ h(u_0))ds + \int_{t_1}^T (g(x(s,t,x_0,u^{t_1}) - g(x(s,t_1,x_0,u^{t_1})))ds. \end{aligned} \tag{7.16}$$

On the other hand, we have

$$\begin{aligned} |x(s,t,x_0,u_0) - x_0| &\leq |e^{(s-t)A}x_0 - x_0| \\ &+ \int_t^{t_1} e^{(s-\theta)A}D^*u_0|d\theta \leq C|t - t_1|(|Ax_0| + 1) \\ &\text{for } t \leq s \leq t_1 \end{aligned}$$

and

$$|x(s,t,x_0,u^{t_1}) - x(s,t_1,x_0,u^{t_1}| \leq |e^{(s-t)A}x_0 - e^{(s-t_1)A}x_0|$$

$$+ |\int_{t_1}^{t} e^{(s-\tau)A}u^{\tau_1}(\tau)d\tau \leq |M|t - t_1|(|Ax_0| + 1)$$

$$\text{for } t_1 \leq s \leq T.$$

Since g is locally Lipschitz we may conclude by (7.16) that for every $x_0 \in D(A)$

$$|\psi(t,x_0) - \psi(t_1,x_0)| \leq L|t - t_1| \quad \text{for } t,t_1 \in [0,T].$$

Next for all $t \in [0,T]$ and $x_0,y_0 \in H$ we have

$$|x(s,t,x_0,u) - x(s,t,y_0,u)| \leq C|x_0-y_0|, \; t \leq s \leq T \tag{7.17}$$

$$|x(s,t,x_0,u)| \leq C(|x_0| + \int_t^T \|u(\tau)\| \, d\tau), \; t \leq s \leq T. \tag{7.18}$$

Let $r > 0$ be arbitrary but fixed. By definition of ψ we see that there exists C_r such that $\psi(t,x_0) \leq C_r$ for $|x_0| \leq r$ and $t \in [0,T]$. According to assumption (7.11) we may restrict to the class V of all $u \in L^2(t,T;U)$ which satisfy the estimate (for some $M_r > 0$)

$$\int_t^T \|u(\tau)\|^2 d\tau \leq M_r$$

and by (7.18), $|x(s,t,x_0,u)| \leq C_r$ for $|x_0| \leq r$.

Recalling that g and ϕ_0 are locally Lipschitz on H we may conclude that for every $r > 0$

$$|\psi(t,x_0) - \psi(t,y_0)| \leq L_r|x_0 - y_0| \text{ for } |x_0|, |y_0| \leq r \tag{7.19}$$

where L_r is independent of t. This completes the proof of Lemma 11.

It is useful to notice that, arguing as in the proof of Proposition 5 in Chapter 1, it follows by (7.19) that

$$\sup \{|W ; W \in \partial\psi(t,x_0)\} \leq C_r \text{ for } |x_0| \leq r. \tag{7.20}$$

For a fixed $(t,x_0) \in [0,T] \times H$ denote by (x^t,u^t) any optimal pair in (7.9), i.e.,

$$\psi(t,x_0) = \int_t^T (g(x^t(s)) + h(u^t(s)))ds + \phi_0(x^t(T)) \tag{7.21}$$

and by $M_t \subset C([t,T],H)$ the set of all $p^t:[t,T] \to H$ which satisfy along with x^t, u^t the system (7.12), (7.13), (7.14). As remarked earlier, $M_t \neq \emptyset$ in virtue of Proposition 11 in Chapter 1.

LEMMA 12. *For every* $(t,x_0) \in [0,T] \times H$, *we have:*

$$\partial\psi(t,x_0) = \{-p^t(t);\ p^t \in M_t\}.$$

Proof Denote by $\mathbb{A}_t$ the set $\{-p^t(t);\ p^t \in M_t\}$. Let $y_0 \in H$ be arbitrary and let (y^t,v^t) be such that

$$\psi(t,y_0) = \int_t^T (g(y^t(s)) + h(v^t(s)))ds.$$

Recalling that

$$(\partial g(x^t), x^t - y^t) \leq g(x^t) - g(y^t) \quad \text{a.e. on }]t,T[$$

and $x^t - y^t$ satisfies (in the mild sense) the system

$$(x^t - y^t)' = A(x^t - y^t) - D^*(u^t - v^t) \text{ on } [t,T]$$

$$(x^t - y^t)(t) = x_0 - y_0$$

we get

$$-(p^t(t), x_0 - y_0) \leq \psi(t,x_0) - \psi(t,y_0).$$

We have therefore proved that $\mathbb{A}_t \subset \partial\psi(t,x_0)$.

To prove that $\mathbb{A}_t = \partial\psi(t,x_0)$ it suffices to show that the operator $x_0 \xrightarrow{\Gamma} \mathbb{A}_t$ is maximal monotone, i.e., $R(I + \Gamma) = H$. For this purpose, for an arbitrary $x_0 \in H$ consider the equation

$$x_0 + \Gamma x_0 \ni z_0 \tag{7.22}$$

which can be equivalently written as

$$\begin{aligned} &(x^t)' = Ax^t - D^*u^t \quad t \leq s \leq T \\ &(p^t)' \in -A^*p^t + \partial g(x^t) \quad t \leq s \leq T \end{aligned} \tag{7.23}$$

$$x^t(t) - p^t(t) = z_0;\ p^t(T) + \partial\phi_0(x^t(T)) \ni 0 \tag{7.24}$$

$$- Dp^t(s) \in \partial h(u^t(s)) \quad \text{a.e. } s \in]t,T[. \qquad (7.25)$$

By Proposition 11 (Chapter 1) the system (7.22)-(7.25) has a solution if and only if the optimization problem

$$\min \{\int_t^T (g(x(s)) + h(u(s)))ds + \frac{1}{2}|x(t) - z_0|^2 + \phi_0(x(T));$$

$$x' = Ax - D^*u,\ t \leq s \leq T;\ u \in L^2(t,T;U)\}$$

does. Since our assumptions imply that the latter has at least one solution we may conclude that Eq. (7.22) has a solution x_0. Thus the proof of Lemma 7 is complete.

Proof of Theorem 9

We shall prove that $\phi(t,x) = \psi(T-t,x)$ satisfies the conditions of Theorem.

Since (7.3), (7.4) follow by Lemmas 11, 12 and equality (7.19) it remains to show that the function $\phi(t,x) = \psi(T-t,x)$ satisfies Eq. (7.6) for $x_0 \in D(A)$ and a.e. $t \in [0,T]$.

Let x_0 be any element of D(A) and let $t \in [0,T]$ be such that $\psi(s,x_0)$ is derivable as a function of s at t. (We notice that almost all t in [0,T] are in this situation). Let (x^t,u^t) be an optimal pair in Problem (7.9), i.e.,

$$\psi(t,x_0) = \int_t^T (g(x^t(s)) + h(u^t(s)))ds + \phi_0(x^t(T)).$$

We must remark that in virtue of the optimality principle (x^t,u^t) is still an optimal pair on every interval [s,T] corresponding to the initial value $x(s) = x^t(s)$. In other words

$$\psi(s,x^t(s)) = \int_s^T (g(x^t(\tau)) + h(u^t(\tau))d\tau + \phi_0(x^t(T)) \qquad (7.26)$$

$$\text{for } t \leq s \leq T.$$

Next we observe that by the optimal system (7.13), (7.14) $u^t(s) = \Phi'(-Dp^t(s))$ is a continuous U-valued function on [t,T] (because $\phi' = \partial h^* = (\partial h)^{-1}$). Then by (7.26) we see that

$$\frac{d}{ds}\psi(s,x^t(s)) + g(x^t(s)) + h(u^t(s)) = 0 \tag{7.27}$$

$$\text{for } s \in [t,T].$$

(We notice that by virtue of Eq. (7.25) and the conjugacy formula the function $s \to h(u^t(s)) = -(Dp^t(s), \partial h^*(-Dp^t(s)) - h^*(-Dp^t(s))$ is continuous.)

By (7.14) and by Lemma 12 we see that:

$$\frac{d}{ds}\psi(s,x^t(s))_{s=t} + g(x_0) + h(\Phi'(D\eta(t,x_0))) = 0 \tag{7.28}$$

where $\eta(t,x_0)$ is an arbitrary element of $\partial\psi(t,x_0)$.

On the other hand, we have

$$\begin{aligned}\frac{d}{ds}\psi(s,x^t(s))_{s=t} &= \lim_{s\to t}\frac{1}{s-t}(\psi(s,x^t(t)) - \psi(t,x^t(t))) \\ &\quad + \lim_{s\to t}\frac{1}{s-t}(\psi(s,x^t(s)) - \psi(s,x^t(t))) \qquad (7.29)\\ &= \psi_t(t,x_0) + \lim_{s\to t}(\psi(s,x^t(s)) - \psi(s,x^t(t)))\cdot\frac{1}{s-t}\,.\end{aligned}$$

Now by the mean value property (see Proposition 6, part (v) in Chapter 1) we have

$$\psi(s,x^t(s)) - \psi(s,x^t(t)) = (\zeta_s,x^t(s) - x^t(t)) \tag{7.30}$$

where $\zeta_s \in \partial\psi(s,z_s)$ and z_s is on the segment $[x^t(t),x^t(s)]$. By (7.19) we see that $\{\zeta_s\}$ is bounded in H as $s \to t$. Thus on some sequence $\{s_n\} \to t$ we have

$$\zeta_{s_n} \to \zeta \quad \text{weakly in H.}$$

Since $z_s \to x^t(t)$ for $s \to t$, letting s_n tend to t in the inequality

$$(\zeta_{s_n},z_{s_n} - v) \geq \psi(s_n,z_{s_n}) - \psi(s_n,v) \quad \forall v \in H$$

we get

$$(\zeta,x^t(t) - v) \geq \psi(t,x^t(t)) - \psi(t,v) \quad \forall v \in H$$

and therefore $\zeta \in \partial\psi(t,x^t(t))$. Since

$$\lim_{s \to t} \frac{x^t(s) - x^t(t)}{s - t} = Ax^t(t) - D^* u^t(t)$$

we see by (7.29), (7.30) and Lemma 12,

$$\frac{d}{ds} \psi(s, x^t(s))_{s=t} = \psi_t(t, x_0) + (\zeta, Ax_0 - D^* \Phi'(D\zeta))$$

and by (7.28) (we may take $\eta(t, x_0) = \zeta$)

$$\psi_t(t, x_0) + (\zeta, Ax_0) + h(\Phi'(D\zeta)) - (D\zeta, \Phi'(D\zeta)) + g(x_0) = 0.$$

Recalling the conjugacy relation (see Proposition 2 in Chapter 2)

$$\Phi^*(\Phi'(D\zeta)) + \Phi(D\zeta) = (D\zeta, \Phi'(D\zeta))$$

we see that (recall that $h = \Phi^*$)

$$\psi_t(t, x_0) - \Phi(D\zeta) + (Ax_0, \zeta) + g(x_0) = 0$$

where $\zeta \in \partial\psi(t, x_0)$. Hence $\phi(t, x) = \psi(T-t, x)$ satisfies Eq. (7.6).

As for uniqueness in Problem (7.6), (7.7) we do not know whether it holds under conditions (7.3), (7.7). However we can prove a weaker uniqueness result in the class $Q \subset B([0,T]; C(H))$ defined below. We shall put in the class Q all functions $\phi: [0,T] \times H \to \mathbb{R}$ which belong to $B([0,T]; C^0_{Lip}(H)) \cap K$, are Lipschitz in t for every $x \in D(A)$ and further satisfy the following conditions:

(*) For every $x \in W^{1,2}([0,T]; H)$ the function $t \to \phi(t, x(t))$ is absolutely continuous on $[0,T]$ and:

$$\frac{d}{dt} \phi(t, x(t)) = \phi_t(t, x(t)) + (\eta(t, x(t)), x'(t)) \qquad (7.31)$$

$$\text{a.e. } t \in [0,T]$$

for all $\eta(t, x) \in \partial\phi(t, x)$.

(**) $u(s) = \Phi'(D\partial\phi(T-s, x(s))$ is a feedback control for the control system

$$x' = Ax - D^* u \qquad s \in [t,T] \qquad (7.32)$$

$$x(t) = x_0.$$

THEOREM 10. *Assume that (a), (b) are satisfied and* A *is the infinitesimal generator of an analytic semigroup. Then the Cauchy Problem (7.6), (7.7) has at most one solution* $\phi \in Q$. *If such a solution* ϕ *exists it is of the form* $\phi(t,x) = \psi(T-t,x_0)$ *where* ψ *is given by Eq. (7.9).*

Proof Let $\phi = Q$ be a solution to (7.6), (7.7) and let $\tilde{\psi}(t,x) = \phi(T-t,x)$. Let $(t,x_0) \in [0,T] \times D(A)$ be arbitrary but fixed and let (x^t,u^t) be an optimal pair in Problem (7.9). Since A generates an analytic semigroup we know that (see Proposition 8 in Chapter 1) $x^t \in W^{1,2}(t,T;H)$. Then by (7.6) and the conjugacy formula we have

$$\frac{d}{dt}\tilde{\phi}(s,x^t(s)) = \tilde{\psi}_s(s,x^t(s)) + (\partial\tilde{\psi}(s,x^t(s)),(x^t)'(s))$$

$$= \tilde{\psi}_s(s,x^t(s)) + (Ax^t(s) - D^*u^t(s)),\partial\tilde{\psi}(s,x^t(s)))$$

$$\geqslant g(x^t(s)) + h(u^t(s)) \quad \text{a.e. } s \in \,]t,T[.$$

Integrating on [t,T] we see that

$$\tilde{\psi}(t,x_0) \leqslant \psi(t,x_0). \tag{7.33}$$

Now let $y = y^t(s)$ be the solution to the closed loop system:

$$\begin{aligned} y' &= Ay - D^*\Phi'(D\partial\phi(T-s,y)), \quad t \leqslant s \leqslant T \\ y(t) &= x_0 \end{aligned} \tag{7.34}$$

which by condition (**) has at least one solution in $W^{1,2}(t,T;H)$. Let y^t be such a solution and let $u^t(s) = \Phi'(D\partial\phi(T-s,y^t(s))$. Again by (7.31) we have

$$\frac{d}{ds}\tilde{\psi}(s,y^t(s)) = \tilde{\psi}_s(s,y^t(s)) + (\partial\tilde{\psi}(s,y^t(s)),$$

$$Ay^t(s) - D^*u^t(s)) = -\,g(y^t(s)) - \Phi^*(u^t(s))$$

which yields

$$\tilde{\psi}(t,x_0) = \int_t^T (g(y^t(s)) + h(u^t(s))ds + \phi_0(y^t(T))$$

$$\geqslant \psi(t,x_0).$$

Along with (7.33) the latter implies $\psi = \tilde{\psi}$ as claimed.

REMARK 8. Another result of the type of Theorem 9 can be found in the book [2] (Chapter IV).

§8 Riccati equation

We shall study the problem:

$$\begin{aligned} &P'(t) - A^*P(t) - P(t)A + P(t)BN^{-1}(t)B^*P(t) = M(t) \\ &P(0) = P_0. \end{aligned} \tag{8.1}$$

Here U is another Hilbert space ($\| \ \|$ norm, $\langle , \rangle$ inner product). We denote by $\mathcal{L}(H)$ the Banach algebra of linear bounded operators $H \to H$ and by $\Sigma(H)$ (resp. $\Sigma^+(H)$) the Banach spaces of the symmetric (resp. positive) operators in H.

Moreover $C_s([0,T]; \Sigma(H))$ is the set of all mappings $S:[0,T] \to \Sigma(H)$ such that $S(\cdot)x$ is continuous for each $x \in H$. By the Banach-Steinhaus Theorem we can endow $C_s([0,T]; \Sigma(H))$ with the norm:

$$\|T\| = \sup \{|T(t)|; \quad t \in [0,T]\}. \tag{8.2}$$

Therefore $C_s([0,T]; \Sigma(H))$ is a closed subspace of $B([0,T]; C(H))$.

We shall assume that

(α) $M,N \in C_s([0,T]; \Sigma^+(H))$, $P_0 \in \Sigma^+(H)$.

(β) $B \in \mathcal{L}(U;H)$.

(γ) There exists $\varepsilon > 0$ such that $N(t) \geqslant \varepsilon$, $\forall t \in [0,T]$.

(δ) A is the generator of a C_0-semigroup e^{tA} on H such that $|e^{tA}| \leqslant Me^{\omega t}$, $M > 0$, $\omega \in \mathbb{R}$.

By a 'mild' solution of Problem (8.1) we mean a function $P \in C_s([0,T]; \Sigma(H))$ such that

$$\begin{aligned} P(t)x = e^{tA^*}P_0\, e^{tA}x + \int_0^t e^{(t-s)A^*}[M(s)-P(s)BN^{-1}(s)B^*P(s)]. \\ e^{(t-s)A}x \, ds, \text{ for all } x \in H. \end{aligned} \tag{8.3}$$

We consider also the approximating problem:

$$\begin{aligned} &P_n'(t) - A_n^*P_n(t) - P_n(t)A_n + P_n(t)BN^{-1}(t)B^*P_n(t) = M(t) \\ &P_n(0) = P_0 \end{aligned} \tag{8.4}$$

which is equivalent to the equation

$$P_n(t)x = e^{tA_n^*}P_0\, e^{tA_n}x + \int_0^t e^{(t-s)A_n^*}[M(s)-P_n(s)BN^{-1}(s)$$

$$B^*P_n(s)]e^{(t-s)A_n}x\ ds,\ \text{for all } x \in H \tag{8.5}$$

where $A_n = n(nI-A)^{-1}-nI$ is the Yosida approximation of A. The main result is:

THEOREM 11. *Under assumptions* (α), (β), (γ), (δ) *there exists a unique solution* P *(resp.* P_n*) to equation (8.3) (resp. (8.4)) which belongs to* $C_s([0,T];\ \Sigma^+(H))$.

Moreover, $P_n(\cdot)x \to P(\cdot)x$ *in* $C([0,T];H)$ $\forall x \in H$.

Proof We prove first, using the contraction principle in $C([0,T];\ \Sigma(H))$ (endowed with norm (8.2)) that there exists a unique local solution to Problems (8.3) and (8.5).

By standard arguments we see that there exist maximal solutions P and P_n defined in intervals $[0,\tau[$ and $[0,\tau_n[$.

Moreover, since $e^{A_nt}x \to e^{At}x$ on every interval we see that for each $0 < T_1 < \tau$,

$$P_n(\cdot)x \to P(\cdot)x \text{ in } C([0,T_1];H) \text{ for (8.6) all } x \in H.$$

Now we show that $P_n(t) \geq 0$; in fact if $t \in [0,T_1]$ we have

$$P_n' - D_nP_n - P_nD_n^* = M,\quad t \in [0,T_1]$$

where $D_n = A_n^* - \frac{1}{2}P_n\ BN^{-1}B^*$.

Call $U_n(t,s)$ the evolution operator attached to $\{D_n(t)\}$; then we have

$$P_n(t)x = U_n^*(t,0)P_0U_n(t,0) + \int_0^t U_n(t,s)M(s)U_n^*(t,s)x\ ds;\quad \forall x \in H \tag{8.6}$$

which implies $P_n(t) \geq 0$ and then, by (8.6), $P(t) \geq 0$. Finally if $x \in H$ and $t \in [0,\tau[$ we have

$$(P_n(t)x,x) = (P_0\, e^{tA}x,e^{tA}x) + \int_0^t \{(M(s)e^{(t-s)A}x,e^{(t-s)A}x) +$$

$$- (N^{-1}(s)B^*P_n(s)e^{(t-s)A}x,B^*P(s)e^{(t-s)A}x)\}ds$$

$$\leq M^2e^{2|\omega|T}|P_0||x|^2 + M^2e^{2|\omega|T}\int_0^t M(s)||x|^2\, ds.$$

It follows that $\sup_{t\in[0,\tau_n[} |P_n(t)| \leq C$, independent of n. This implies $\tau_n = \tau = T$ and

$$P_n(\cdot)x \to P(\cdot)x \text{ in } C([0,T];H) \tag{8.7}$$

and moreover $P(t) \geq 0$, $\forall t \in [0,T]$.

We give now an application to the quadratic cost control problem (see §4 in Chapter 1).

PROPOSITION 3. *Assume that the hypotheses of Theorem 11 are fulfilled. Set* $\phi(t,x) = \frac{1}{2}(P(t)x,x)$. *Then if* $x \in D(A)$ *we have:*

$$\phi_t(t,x) + \frac{1}{2}(BN^{-1}(t)B^*(\phi)_x(t,x),\ (\phi)_x(t,x))$$

$$- (A\,x,(\phi)_x(t,x)) = \frac{1}{2}(M(t)x,x) \tag{8.8}$$

$$\phi(0,x) = \frac{1}{2}(P_0x,x).$$

Moreover if $\psi(t,x) = \frac{1}{2}(P(T-t)x,x)$, *we have*

$$\psi(t,x_0) = \min \{\frac{1}{2}\int_t^T [(M(s)x(s),x(s)) + (N(s)u(s),u(s))]ds$$
$$+ \frac{1}{2}(P_0x(T),x(T));\ x' = Ax + Bu, x(t) = x\} \tag{8.9}$$

and the optimal control u^* *is given by*

$$u^*(s) = -N^{-1}B^*P(T-s)y^*(s) \tag{8.10}$$

where y^* *is the solution of the closed loop differential equation*

$$\begin{cases} y' = (A - BN^{-1}B^*P(T-s))y \\ y(t) = x_0 \end{cases} \tag{8.11}$$

Proof It is easy to see that

$$\phi_n' + \frac{1}{2}(BN^{-1}B^*(\phi_n)_x,(\phi_n)_x) - (A_n x,(\phi_n)_x) = \frac{1}{2}(Mx,x)$$

$$\phi_n(0,x) = \frac{1}{2}(P_0 x,x)$$

where $\phi_n(t,x) = \frac{1}{2}(P_n(t)x,x)$. Moreover, by (8.7) we see that

$$\phi_n(t,x) \to \phi(t,x),\ (\phi_n)_x(t,x) \to \phi(t,x),\ \text{unif. in } t.$$

If $x \in D(A)$ then it follows that

$$\phi_n'(t,x) \to \frac{1}{2}(M(t)x,x) - \frac{1}{2}(BN^{-1}(t)B^*(\phi_n)_x(t,x),(\phi_n)_x(t,x))$$
$$- (Ax,(\phi_n)_x(t,x)) \text{ unif. in } [0,T]$$

and Eq. (8.8) follows.

The proof of other assertions is completely similar to that of Proposition 2, so we shall omit it.

We will now study the existence of a solution to Problem (8.1). We set

$$\mathbb{A}(S)x = \lim_{t\downarrow 0} \frac{1}{t}(e^{tA^*} S e^{tA}x - Sx),\quad S \in D(\mathbb{A})$$
$$D(\mathbb{A}) = \{S \in \Sigma(H);\ \lim_{t\downarrow 0} \frac{1}{t}(e^{tA^*} S e^{tA}x - Sx) \in \Sigma(H)\} \tag{8.12}$$

($\mathbb{A}$) is the restriction to $\Sigma(H)$ of the operator defined by (5.6), (5.7)).

If $S \in D(\mathbb{A})$ and $x \in D(A)$ then $Sx \in D(A^*)$ and

$$\mathbb{A}(S)x = A^*Sx + SAx \tag{8.13}$$

(see Lemma 10).

Now we may write Eq. (8.1) in terms of $\mathbb{A}$, as

$$P' - \mathbb{A}(P) + PBN^{-1}B^*P = M$$
$$P(0) = P_0. \tag{8.14}$$

We say that P is a *strong solution* to Eq. (8.1) if

$$\begin{aligned}
&\text{(a)} \quad P \in C_s([0,T];\ \Sigma(H)) \cap C_s([0,T];D(\mathbb{A})) \\
&\text{(b)} \quad P(\cdot)x \text{ is differentiable } \forall x \in H \text{ and} \\
&\qquad \frac{dP}{dt} \in C_s([0,T];\ \Sigma(H)) \\
&\text{(c)} \quad P \text{ fulfils (8.14).}
\end{aligned} \tag{8.15}$$

If P is a strong solution then for any $x \in D(A)$ we have $P(t)x \in D(A^*)$ and Eq. (8.1) holds.

PROPOSITION 4. *Assume that the hypotheses of Theorem 11 are fulfilled and in addition that* $M_t \in C_s([0,T];\ \Sigma(H))$ *and* $P_0 \in D(\mathbb{A})$. *Then Problem (8.1) has a unique strong solution.*

Proof Consider the Cauchy problem:

$$\begin{aligned}
&P_n' - \mathbb{A}_n(P_n) + P_nBN^{-1}B^*P_n = M \\
&P_n(0) = P_0
\end{aligned} \tag{8.16}$$

where $\mathbb{A}_n(T)x = n^2 \int_0^\infty e^{-nt} e^{tA^*} T\, e^{tA} x\, dt - n\, Tx$; proceeding as in the proof of Theorem 11 we see that

$$P_n(\cdot)x \to P(\cdot)x \quad \forall x \in H \text{ in } C([0,T]);H). \tag{8.17}$$

Consider also the equations

$$\begin{aligned}
&V' - \mathbb{A}(V) + VBN^{-1}B^*P + PBN^{-1}B^*V = M' \\
&V(0) = \mathbb{A}(P_0) + P_0BN^{-1}B^*P_0 + M(0)
\end{aligned} \tag{8.18}$$

and

$$\begin{aligned}
&V_n' - \mathbb{A}_n(V_n) + V_nBN^{-1}B^*P_n + P_nBN^{-1}B^*V_n = M' \\
&V_n(0) = V(0).
\end{aligned} \tag{8.19}$$

Equations (8.18) and (8.19) are obtained by formal derivation in t of Problems (8.14) and (8.16).

It is easy to see that (8.18) and (8.19) have a unique solution and moreover

$$P_n'(\cdot)x = V_n(\cdot)x \to V(\cdot)x \text{ in } C([0,T];H), \quad \forall x \in H \tag{8.20}$$

which along with (8.17) gives $V = P'$; and the claim follows.

We conclude this chapter with some additional regularity results in the case where the semigroup e^{tA} is analytic. We assume here:

$$|A\, e^{tA}| < \frac{K}{t} \quad \forall t > 0, \; G_t(T) = e^{tA^*}\, T\, e^{tA} \tag{8.21}$$

which implies

$$|AG_t(T)| = \left|\frac{dG_t(T)}{dt}\right| < \frac{2K}{t}|T|, \quad \forall t \in \Sigma(H). \tag{8.22}$$

We say that $P:[0,T] \to \Sigma(H)$ is a *classical solution* to Problem (8.1) if:

(a') P is continuously differentiable (in the topology of $\mathcal{L}(E)$) in $]0,T]$

(b') $P(t) \in D(A)$ $\forall t \in]0,T]$ and AP is continuous in $]0,T]$ (in the topology of $\mathcal{L}(E)$) (8.23)

(c') P fulfils the equations (8.14).

To prove the existence of a classical solution we need the following lemma.

LEMMA 13. *Let* $V \in C_s([0,T];\, \Sigma(H))$ *and let* F *be defined by:*

$$F(t)\cdot x = \int_0^t G_{t-s}(V(s))\cdot x \, ds \quad \forall x \in H. \tag{8.24}$$

Then $F \in C^\alpha([0,T];\, \Sigma(H))$, $\forall \alpha \in]0,1[$.
If moreover $V \in C^\alpha([0,T];\, \Sigma(H))$, *then we have:*

$$F \in C^1([0,T];\, \Sigma(H)) \cap C([0,T];D(A)).$$

If finally $V \in C^k([0,T];\, \Sigma(H))$, *then for any* $\varepsilon > 0$ *we have* $F \in C^k([\varepsilon,T];\Sigma(H))$.

Proof If $t > \tau > 0$, we have

$$F(t) - F(\tau) = \int_\tau^t G_{t-s}(V(s))\, ds + \int_0^\tau ds \int_{\tau-s}^{t-s} AG\, (V(\sigma))d\sigma.$$

Recalling (8.21) and (8.22) it follows that

$$|F(t) - F(\tau)| \leq C\left\{|t - \tau| + \int_0^\tau ds \int_{\tau-s}^{t-s} \frac{d\sigma}{\sigma}\right\}$$

$$\leq C_\alpha |t - \tau|^\alpha \qquad \forall \alpha \in]0,1[$$

so the first assertion follows. As regards the second, if $V \in C^\alpha([0,T];\Sigma(H))$ it is easy to prove (using the Yosida approximating A_n of A) the formula:

$$F'(t)x = \int_0^t AG_{t-s}(V(s) - V(t))x\, ds \qquad (8.25)$$
$$+ G_t(V(t))x, \forall x \in H.$$

Note that formula (8.25) is meaningful because

$$\|AG_{t-s}(V(s) - V(t))\| \leq \frac{C}{(t-s)^{1-\alpha}} \cdot$$

From (8.25) it follows that $F \in C^1([0,T]; \Sigma(H)) \cap C([0,T];D(A))$ as claimed.

Finally if $V \in C^k([0,T]; \Sigma(H))$ and $t > 0$ it is easy to see that

$$F^{(k)}(t)x = A^{k-1}G_t(V(0))x + A^{k-2}G_t(V'(0))x + \ldots +$$
$$G_t(V^{(k-1)}(0))x + \int_0^t G_s(V^{(k)}(t-s))x\, ds$$

so the last claim follows.

PROPOSITION 5. *Assume that the hypotheses of Theorem 11 are fulfilled as well as hypothesis (8.21), and let* P *be the solution to Eq. (8.3). Then for any* $\varepsilon \in]0,T[$ *we have* $P \in C^\alpha([\varepsilon,T]; \ (H))$. *Moreover if,* $M \in C^\alpha([0,T]; \Sigma(H))$ *then* P *is a classical solution to Problem (8.1).*

If finally $M \in C^k([0,T], \Sigma(H))$, $k = 1,2,\ldots,$ *then we have* $P \in C^k([\varepsilon,T];\Sigma(H))$ *for any* $\varepsilon \in]0,T[$.

Proof Let $P \in C_s([0,T]; \Sigma^+(H))$ be the solution of Eq. (8.3). Set $V = PBN^{-1}B^*P + M$, then we have

$$P(t)x = G_t(P_0) + \int_0^t G_{t-s}(V(s))x\, ds \quad \forall x \in H.$$

The first term in the right-hand side is analytic for $t > 0$ and the second belongs to $C^\alpha([0,T]; \Sigma(H))$ in virtue of Lemma 13. Then for any $\varepsilon \in]0,T[$ we have $P \in C^\alpha([\varepsilon,T]; \Sigma(H))$. Assume now $M \in C^\alpha([0,T]; \Sigma(H))$; then we have

$$V \in C^\alpha([\varepsilon,T]; \Sigma(H)) \text{ for any } \varepsilon \in]0,T[.$$

For any $t \in [\varepsilon,T]$ we have

$$P(t)x = G_{t-\varepsilon}(P(\varepsilon))x + \int_\varepsilon^t G_{t-s}(V(s))x\, ds \quad \forall x \in H$$

from which, using once again Lemma 13, we get

$$P \in C^1([2\varepsilon,T]; \Sigma(H)) \cap C([2\varepsilon,T];D(A)).$$

Due to the arbitrariness of ε it follows that P is a classical solution of Problem (8.1).

The last claim is an easy consequence of Lemma 13.

REMARK 9. It can be shown that, under the hypotheses of Proposition 5 we have $P \in C^{1,\alpha}([\varepsilon,T]; \Sigma(H)) \cap C^\alpha([\varepsilon,T];D(A))$ for any $\varepsilon \in]0,T[$ (see [16]).

REMARK 10. Assume that M is independent of t, then from Proposition 5 it follows $P \in C^\infty([\varepsilon,T]; \Sigma(H))$ for any $\varepsilon > 0$.

COROLLARY 1. *Assume that the hypotheses of Proposition 5 hold, and moreover that* M *is constant.*

Let P *be the solution to Eq. (8.3),* $y^* \in C([0,T];H))$ *the solution of the closed loop equation (8.11) and* u^* *given by (8.10). Then for any* $\varepsilon \in]0,T[$ *we have* $y^* \in C^\infty([0,T-\varepsilon];H)$ *and* $u^* \in C^\infty([0,T-\varepsilon];H)$.

Proof It is an easy consequence of Remark 10 and of well-known properties of abstract linear equations.

§9 Stationary Hamilton-Jacobi equations

We shall study here the operator equation

$$\frac{1}{2}|\phi'(x)|^2 - (Ax,\phi'(x)) = g(x), \quad \forall x \in D(A) \tag{9.1}$$

in the real Hilbert space H under the following assumptions:

(j) *The linear operator* A *is the infinitesimal generator of a* C_0*-semigroup on* H. *There exists* $\omega \in \mathbb{R}$ *such that*

$$(Ax,x) \leq \omega\, |x|^2, \text{ for all } x \in D(A) \tag{9.2}$$

(jj) *The function* $g:H \to \mathbb{R}$ *belongs to* K *and satisfies the following two conditions.*

$$\textit{If } g(x_n^0) \to 0 \textit{ for some sequences } \{x_n^0\} \textit{ then } x_n^0 \to 0 \tag{9.3}$$

$$g(\exp(-(\lambda I-A)t)x_0 \in L^1(\mathbb{R}^+) \textit{ for } \lambda > \omega \textit{ and all } x_0 \in H. \tag{9.4}$$

Here $\exp(-(\lambda I-A)t)$ is the semigroup generated by the operator $-\lambda I + A$ and K is the closed convex cone of C(H) consisting of all convex functions $\phi:H \to \mathbb{R}$ which satisfy the condition: $\phi(x) \geq \phi(0) = 0$ for all $x \in H$.

By a solution to (9.1) we shall mean a function $\phi:H \to \mathbb{R}$ which is Gâteaux differentiable and satisfies Eq. (9.1) for all $x \in D(A)$.

Equation (9.1) is relevant in the calculus of variations with infinite horizon and in nonlinear analysis. The main result is:

THEOREM 12. *Let assumptions* (j) *and* (jj) *part (9.4) be satisfied. Then Eq. (9.1) has at least one solution* $\phi^\infty \in K$. *If also condition (9.3) holds then the solution* ϕ^∞ *to Eq. (9.1) is unique and for every sequence* $\{x_n^0\}$ *having the property:* $\lim_{n\to\infty} \phi^\infty(x_n^0) = 0$ *we have:*

$$\lim_{n\to\infty} e^{tA}x_n^0 = 0 \ \forall t > 0.$$

Proof Define the function $\phi^\infty:H \to \mathbb{R}$,

$$\phi^\infty(x_0) = \inf \{\int_0^\infty (g(x(t)) + \frac{1}{2}|u(t)|^2)dt;\ x' = Ax + u,$$
$$x(0) = x_0;\ u \in L^2(\mathbb{R}^+;H)\}. \tag{9.5}$$

By (9.4) we see that $\phi^\infty < +\infty$ on H and for every $x_0 \in H$ the infimum defining $\phi^\infty(x_0)$ is attained in a unique pair (x^*,u^*). Moreover, the function ϕ^∞ is convex and l.s.c. on H. Since it is everywhere finite we conclude that it is continuous on H. Finally arguing as in §7 it follows that for every $t \geq 0$, (x^*,u^*) is an optimal pair for the finite horizon control problem:

$$\phi^\infty(x_0) = \inf \{\int_0^t (g(x(s)) + \frac{1}{2}|u(s)|^2)ds + \phi^\infty(x(t));$$

$$x' = Ax + u,\ x(0) = x_0\} \tag{9.6}$$

$$= \int_0^t (g(x^*(s)) + \frac{1}{2}|u^*(s)|^2)ds + \phi^\infty(x^*(t)).$$

Thus according to Proposition 11 in Chapter 1, for every $t \geq 0$ there exists $p^t \in C([0,T];H)$ which satisfies the system

$$(p^t)' = - A^*p^t + \partial g(x^*), \quad 0 \leq s \leq t \tag{9.7}$$

$$p^t(t) + \partial\phi^\infty(x^*(t)) \ni 0 \tag{9.8}$$

$$p^t(s) = u^*(s) \qquad s \in [0,T] \tag{9.9}$$

Hence u^* is the solution to

$$(u^*)' \in -A^*u^* + \partial g(x^*) \qquad t \in \mathbb{R}^+ \tag{9.10}$$

$$u^*(t) + \partial\phi^\infty(x^*(t)) \ni 0 \quad \forall t \geq 0 \tag{9.11}$$

(Eq. (9.10) is understood of course in the mild sense). Next by (9.6) we see that

$$\phi^\infty(x^*(t)) = \int_t^\infty (g(x^*(s)) + \frac{1}{2}|u^*(s)|^2)ds \quad \forall t \geq 0. \tag{9.12}$$

Assume that $x_0 \in D(A)$. Then x^* is right differentiable at $t = 0$ and $\frac{d^+}{dt} x^*(0) = Ax_0 + u^*(0)$. Since

$$t^{-1}(\phi^\infty(x^*(t)) - \phi^\infty(x_0)) \xrightarrow{t \to 0} (\eta, \frac{d^+}{dt} x^*(0))$$

where η is an element of $\partial\phi^\infty(x_0)$ it follows by (9.12) that

$$(\eta, Ax_0 + u^*(0)) + g(x_0) + \frac{1}{2}|u^*(0)|^2 = 0 \tag{9.13}$$

where $-u^*(0) \in \partial\phi^\infty(x_0)$ and $\eta \in \partial\phi^\infty$. To conclude the proof of existence it suffices to show that ϕ^∞ is Gâteaux differentiable, i.e., $\partial\phi^\infty$ is single valued.

To this end we define the operator $\Gamma : H \to H$,

$$\Gamma x_0 = - p(0)$$

where $p \in C([0,T];H)$ is the solution to the system (T is fixed)

$$x' - Ax = p \qquad 0 \leqslant t \leqslant T$$

$$p' + A^* p \in g(x) \qquad 0 \leqslant t \leqslant T \tag{9.14}$$

$$x(0) = x_0, \; p(T) + \partial\phi^{\infty}(x(T)) \ni 0.$$

By Proposition 11 (Chapter 1) (x,p) is a solution to (9.14) if and only if (x,p) is an optimal pair for the problem

$$\inf \{\int_0^T (g(x) + \tfrac{1}{2}|p|^2)dt + \phi(x(T)); \; x' = Ax + u, \; x(0) = x_0\}. \tag{9.15}$$

Since the solution of the latter problem is unique we conclude that Γ is single valued. On the other hand, by (9.11) we see that $\Gamma x_0 \in \partial\phi^{\infty}(x_0)$. To prove that $\Gamma = \partial\phi^{\infty}$ it suffices to show that the range $R(I + \Gamma)$ is all of H. To this end we consider the equation $x_0 + \Gamma x_0 = y_0$ which is equivalent to

$$x' = Ax + p \qquad 0 \leqslant t \leqslant T$$

$$p' = A^* p + \partial g(x) \qquad 0 \leqslant t \leqslant T \tag{9.16}$$

$$x(0) - p(0) = y_0, \; p(T) + \partial\phi^{\infty}(x(T)) \ni 0$$

and again by Proposition 11 it is equivalent to the control problem:

$$\inf \{\int_0^T (\tfrac{1}{2}|u|^2 + g(x))dt + \phi(x(T)) + \tfrac{1}{2}|x(0) - y_0|^2; \quad x' + Ax = u\} \tag{9.17}$$

which clearly admits at least one solution $(x, p = u)$ which is also a solution to (9.16).

Hence $\Gamma = \partial\phi^{\infty}$, and $\partial\phi^{\infty} = (\phi^{\infty})$ is single valued as claimed. As for uniqueness we consider the differential equation

$$\begin{aligned} &x' = Ax - \phi_x^0(x) \qquad t \geqslant 0 \\ &x(0) = x_0 \end{aligned} \tag{9.18}$$

where ϕ^0 is any solution to Eq. (9.1).

Since $\omega I - A + \phi_x^0$ is m-accretive (because $\omega I - A$ is m-accretive and ϕ_x^0 is monotone and demicontinuous on H) for each $x_0 \in D(A)$, Eq. (9.18) has a unique solution $\tilde{x} \in W^{1,\infty}_{loc}(\mathbb{R}^+;H)$. We set $\tilde{u} = -\phi_x^0(x)$ and take the inner product of (9.18) by $\tilde{u}$. Since ϕ^0 is a solution to Eq. (9.1) we get

$$\frac{d}{dt}\,\phi^0(\tilde{x}(t)) + \frac{1}{2}|\tilde{u}(t)|^2 + g(\tilde{x}(t)) = 0 \quad \text{a.e.} \quad t > 0$$

and therefore

$$\phi^0(x_0) = \phi^0(\tilde{x}(t)) + \int_0^t (g(\tilde{x}(s)) + \frac{1}{2}|\tilde{u}(s)|^2)ds \quad \forall t > 0.$$

Finally

$$\phi^0(x_0) = \int_0^\infty (g(\tilde{x}(s)) + \frac{1}{2}|\tilde{u}(s)|^2)ds > \phi^\infty(x_0) \tag{9.19}$$

where ϕ^∞ is given by (9.5). (To get (9.19) we have used the fact that $\tilde{x}(t_n) \to 0$ for some $t_n \to \infty$ because $g(\tilde{x}) \in L^1(\mathbb{R}^+)$ and g satisfies condition (9.3).)

Now let (x^*,u^*) be the optimal pair for the Problem (9.5) $(x_0 \in D(A))$. Again by Eq. (9.1) we have

$$\frac{d}{dt}\,\phi^0(x^*(t)) = (x^{*\prime}(t),\phi_x^0(x^*(t))) = (Ax^*(t) + u^*(t),$$

$$\phi_x^0(x^*(t))) = (u^*(t),\phi_x^0(x^*(t))) - g(x^*(t))$$

$$+ \frac{1}{2}|\phi_x^0(x^*(t))|^2 > - g(x^*(t)) - \frac{1}{2}|u^*(t)|^2, \quad \text{a.e. } t > 0$$

and therefore

$$\phi^0(x^*(t)) + \int_0^t (\frac{1}{2}|u^*(s)|^2 + g(x^*(s)))ds > \phi^0(x_0) \quad \forall t > 0.$$

Since $g(x^*) \in L^1(\mathbb{R}^+)$ it follows that $\lim_{t_n\to\infty} x^*(t_n) \to 0$ for some sequence $t_n \to \infty$ and therefore

$$\lim_{t_n\to\infty} \phi_\infty^0(x^*(t_n)) = \phi^0(0) = 0.$$

Hence $\phi^0(x_0) < \phi^\infty(x_0)$ for all $x_0 \in D(A)$. Along with (9.19) the latter implies that $\phi^{\infty} = \phi^0$ as claimed.

Now let $\{x_n^0\} \subset H$ be such that $\phi^{\infty}(x_n^0) \to 0$ for $n \to \infty$. Let (x_n^*, u_n^*) be an optimal pair in (9.5) where $x_0 = x_n^0$, i.e.,

$$\phi^{\infty}(x_n^0) = \int_0^{\infty} (g(x_n^*) + \frac{1}{2}|u_n^*|^2)dt$$

and

$$x_n^{*\prime} = A\, x_n^* + u_n^* \text{ for } t \geqslant 0;\ x_n^*(0) = x_n^0.$$

Clearly $u_n^* \to 0$ in $L^2(\mathbb{R}^+;H)$ and

$$x_n^*(t) \to 0 \quad \text{a.e. } t > 0 \text{ for } n \to \infty.$$

Now by the variation of constant formula

$$|x_n^*(t) - e^{tA}x_n^0| \leqslant \int_0^t |e^{(t-s)A}u_n^*(s)|ds \leqslant \varepsilon_n \text{ for } 0 \leqslant t \leqslant 1$$

where $\varepsilon_n \to 0$ for $n \to \infty$.

Hence $e^{tA}x_n^0 \to 0$ a.e. $t > 0$ and this implies that $e^{tA}x_n^0 \to 0$ for all $t > 0$ as claimed.

Now let $g, \phi_0 \in C^1_{Lip}(H)$ and let ϕ be the weak solution to the Cauchy Problem (4.1) in $[0, +\infty[$.

The existence and uniqueness of such a solution has been proved by Theorem 3.

Now we shall prove the following asymptotic result:

THEOREM 13. *Let* A *be the infinitesimal generator of a* C_0*-semigroup of contractions on* H*; let* $g \in K \cap C^1_{Lip}(H)$ *satisfy assumptions* (jj). *Then for every* $\phi_0 \in K \cap C^1_{Lip}(H)$ *the solution* ϕ *to the problem:*

$$\begin{aligned} &\phi_t + \frac{1}{2}|\phi_x|^2 - (Ax, \phi_x) = g;\ t \geqslant 0,\ x \in D(A) \\ &\phi(0,x) = \phi_0(x), \quad x \in H \end{aligned} \tag{9.20}$$

satisfies

$$\lim_{t\to\infty} \phi(t,x) = \phi^{\infty}(x) \text{ for all } x \in H. \tag{9.21}$$

Proof By Remark 7, ϕ is given by Eq. (6.15). This yields

$$\int_0^t (g(x^t(s)) + \frac{1}{2}|u^t(s)|^2)ds + \phi_0(x^t(t))$$
$$= \phi(t,x_0) \leq \int_0^t (g(x^*(s)) + \frac{1}{2}|u^*(s))|^2)ds + \phi_0(x^*(t)) \tag{9.22}$$

where (x^*,u^*) is the optimal pair in Problem (9.5) and (x^t,u^t) in (6.5), i.e.,

$$(x^t)' = Ax^t + u^t \text{ on } [0,t]$$
$$x^t(0) = x_0. \tag{9.23}$$

Letting $t \to +\infty$ we see by (9.22) that

$$\limsup_{t \to \infty} \phi(t,x_0) \leq \phi^\infty(x_0). \tag{9.24}$$

We set

$$\tilde{u}^t(s) = \begin{cases} u^t(s) & \text{for } 0 \leq s \leq t \\ 0 & \text{for } s > t \end{cases}$$

and

$$\tilde{x}^t(s) = e^{sA}x_0 + \int_0^s e^{(s-\tau)A}\tilde{u}^t(\tau)d\tau.$$

By (9.22) we see that on some sequence $t_n \to \infty$, $\tilde{u}^{t_n} \to \tilde{u}$ weakly in $L^2(\mathbb{R}^+;H)$ and $\tilde{x}^t(s) \to \tilde{x}(s)$ weakly in H for every $s \geq 0$, where $\tilde{x}$ is the mild solution to

$$x' = Ax + \tilde{u} \text{ on } \mathbb{R}^+$$
$$x(0) = x_0. \tag{9.25}$$

Since the functional

$$(y,u) \to \int_0^T (g(y) + \frac{1}{2}|u|^2)dt$$

is weakly lower semicontinuous on every $L^2(0,T; \times H) \times L^2(0,T;H)$ (because it is convex and lower semicontinuous) it follows by (9.22) and (9.24) that

$$\int_0^\infty (g(\tilde{x}(s)) + \frac{1}{2}|\tilde{u}(s)|^2)ds \le \limsup_{t\to\infty} \phi(t,x_0) \le \phi^\infty(x_0)$$

which obviously implies (9.21) as claimed.

Now we shall develop an iterative procedure to solving Eq. (9.1). First we shall exhibit some simple properties of the equation

$$\lambda\phi + \frac{1}{2}|\phi_x|^2 - (Ax,\phi_x) = g \tag{9.26}$$

where $\lambda > 0$ and A is m-dissipative. We shall consider the approximating equation

$$\lambda\phi^\varepsilon + \varepsilon^{-1}(\phi^\varepsilon - S(\varepsilon)\phi^\varepsilon) - (Ax,\phi_x^\varepsilon) = g \tag{9.27}$$

where $S(\varepsilon)$ has been defined in §2 by formula (2.11). We shall write Eq. (9.27) in the following 'mild' form

$$\phi^\varepsilon(x) = \int_0^\infty e^{-(1+\lambda\varepsilon)t}((S(\varepsilon)\phi^\varepsilon)(e^{\varepsilon tA}x) + \varepsilon g(e^{\varepsilon tA}x)dt = \Gamma_\varepsilon(\phi^\varepsilon)(x). \tag{9.28}$$

Invoking the calculations made in §2, 3, 4 we see that

$$|\Gamma_\varepsilon(\phi)|_{0,r} \le (1+\lambda\varepsilon)^{-1}(|\phi|_{0,r} + \varepsilon|g|_{0,r}), \quad \forall\phi \in K \tag{9.29}$$

$$|\Gamma_\varepsilon(\phi) - \Gamma_\varepsilon(\psi)|_{0,r} \le (1+\lambda\varepsilon)^{-1}|\phi-\psi|_{0,r}, \quad \forall\phi, \psi \in K. \tag{9.30}$$

Hence for every $\varepsilon > 0$, Γ_ε maps K into itself and it is a contraction in $C(\Sigma_r)$. Hence for every $\varepsilon > 0$, Eq. (9.28) has a unique solution $\phi_\varepsilon \in K$.

In Lemma 14 below we gather for later use some elementary properties of ϕ_ε.

LEMMA 14. *Assume that* $g \in K \cap C^1_{Lip}(H)$ *and let* ϕ^ε *be the solution Eq. (9.27). Then the following estimates hold.*

$$|\phi^\varepsilon|_{0,r} \le \lambda^{-1}|g|_{0,r}; \ |\phi^\varepsilon|_{1,r} \le \lambda^{-1}|g|_{1,r} \tag{9.31}$$

$$\|\phi^\varepsilon\|_{1,r} \le (1-C_r\varepsilon)(\lambda-C_r(1+\lambda\varepsilon))^{-1}\|g\|_{1,r} \tag{9.32}$$

$$|\phi_\varepsilon-\tilde{\phi}_\varepsilon|_{0,r} \le \lambda^{-1}|g-\tilde{g}|_{0,r} \tag{9.33}$$

where $\tilde{\phi}_\varepsilon$ *is the solution corresponding to* $\tilde{g}$. *Moreover, we have*

$$\lambda\phi^\varepsilon(x) + \frac{1}{2}|\phi_x^\varepsilon(x)|^2 - (Ax,\phi_x^\varepsilon(x)) = g(x) + R_\varepsilon(x) \quad \text{for all } x \in D(A) \tag{9.34}$$

where

$$|R_\varepsilon|_{0,r} \leqslant M_r\varepsilon \text{ for all } r > 0 \tag{9.35}$$

Proof Estimates (9.31)-(9.33) follow immediately from (9.28) and the corresponding properties of $S(\varepsilon)$. As regards (9.35) it follows by (9.31) and (9.32).

We may now formulate an existence result for Eq. (9.26).

PROPOSITION 6. *For every* $\lambda > 0$ *and any* $g \in K \cap C^1_{Lip}(H)$, *Eq. (9.26) has a unique weak solution* $\phi \in K$ *in the following sense: there exists a sequence* $\{\phi^\varepsilon\} \subset K \cap C^1_{Lip}(H)$ *such that* $\phi^\varepsilon \to \phi$ *in* $C(H)$ *and*

$$\lambda\phi^\varepsilon + \frac{1}{2}|\phi_x^\varepsilon|^2 - (Ax,\phi_x^\varepsilon) \to g \text{ in } C(H) \text{ for } \varepsilon \to 0 \tag{9.36}$$

Proof Let ϕ^ε be the solution to Eq. (9.27). By Lemma 14 we have for all $\varepsilon, \eta > 0$

$$\lambda\phi^\eta + \varepsilon^{-1}(\phi^\eta - S(\varepsilon)\phi^\eta) - (Ax,\phi_x^\eta) = g + R_\varepsilon - R_\eta$$

where $|R_\varepsilon|_{0,r} \leqslant \varepsilon$, $|R_\eta|_{0,r} \leqslant \eta$. Then by (9.33) it follows that

$$|\phi_\varepsilon - \phi_\varepsilon|_{0,r} \leqslant |R_\varepsilon - R_\eta|_{0,r} \leqslant M_r|\varepsilon - \eta|$$

where M_r is independent of ε and η. Hence $\phi = \lim_{\varepsilon\to 0} \phi^\varepsilon$ exists in the topology of $C(H)$.

By (9.34) we see that also (9.36) holds. Hence ϕ is a weak solution to Eq. (9.26).

Now let $\Gamma:K \to K$ be the operator defined by $\psi = \Gamma\phi$ where $\psi \in K$ is the solution to the equation

$$\psi + \frac{1}{2}|\psi_x|^2 - (Ax,\psi_x) = \phi.$$

By Proposition 6, Γ is well defined and if $\phi \leqslant \tilde{\phi}$ then $\Gamma\phi \leqslant \Gamma\tilde{\phi}$. The latter follows by Proposition 3 observing that if $g \leqslant \tilde{g}$ on H then the corresponding solutions ϕ_ε and $\tilde{\phi}_\varepsilon$ to Eq. (9.28) are in the relation: $\phi_\varepsilon \leqslant \tilde{\phi}_\varepsilon$.

Let $\{\phi_n\} \subset K$ be the sequence defined by

$$\phi_{n+1} = \Gamma(\phi_n + g); \quad n = 1,2,\ldots \tag{9.38}$$

where $\phi_0 \in K$ is arbitrary.

One may speculate that $\{\phi_n\}$ is convergent for $n \to \infty$ to the solution ϕ^∞ of Eq. (9.1) provided by Theorem 12. We shall see this is indeed the case under some additional assumptions.

THEOREM 14. *Let* A *be the infinitesimal generator of a* C_0*-semigroup of contractions on* H *and let* $g \in C^1_{Lip}(H) \cap K$ *satisfy assumptions* (jj). *Then there exists* $\phi \in K$ *such that*

$$\lim_{n\to\infty} \phi_n(x) = \phi(x) \qquad \forall x \in H. \tag{9.39}$$

Assume in addition that the space H *has a countable orthonormal basis consisting of eigenfunctions of* A. *Then* $\phi = \phi^\infty$.

Proof As remarked earlier the sequence $\{\phi_n\}$ is monotonically increasing for every $x \in H$. On the other hand, for every $r > 0$ there exists $\tilde{\phi} \in K \cap C^1(H)$ such that

$$\frac{1}{2}|\tilde{\phi}_x(x)|^2 - (Ax,\tilde{\phi}_x(x)) \geqslant g(x) \text{ for } |x| \leqslant r$$

(For instance, we may choose $\phi(x) = \mu|x|^{3/2}$ where $\mu > 0$.) This yields,

$$\phi_n(x) \leqslant \tilde{\phi}(x) \quad \text{for all } x \in \Sigma_r$$

and therefore $\phi(x) = \lim_{n\to\infty} \phi_n(x)$ exists for all $x \in H$. Obviously $\phi \in K$. It remains to show that under the supplementary assumptions we imposed this function is just the solution ϕ^∞ to Eq. (9.1). As an intermediate step we shall prove the following lemma.

LEMMA 15. *Let* H *be a finite dimensional Hilbert space. Then* $\phi = \phi^\infty$.

Proof Let x_n^ε be the solution to differential system

$$\{x_n^\varepsilon\}' = Ax_n^\varepsilon - (\phi_{n+1}^\varepsilon)_x(x_n^\varepsilon) \text{ on } \mathbb{R}^+$$
$$x_n^\varepsilon(0) = x_0 \tag{9.40}$$

where $\{\phi_{n+1}^\varepsilon$ is the sequence associated with $\{\phi_{n+1}\}$ by Proposition 6, i.e.,

$$\phi_{n+1}^\varepsilon(x) + \frac{1}{2}|(\phi_{n+1}^\varepsilon)_x(x)|^2 - (Ax,(\phi_{n+1}^\varepsilon)_x(x))$$
$$= g(x) + \phi_n^\varepsilon(x) + R_n^\varepsilon(x) \tag{9.41}$$

where $|R_n^\varepsilon(x)| \leq M_r\varepsilon$ for $|x| \leq \varepsilon$.

We multiply Eq. (9.40) by $(\phi_{n+1}^\varepsilon)_x(x_n^\varepsilon)$ and integrate on $[0,t]$ to get

$$\phi_{n+1}^\varepsilon(x_0) = \phi_{n+1}^\varepsilon(x_n^\varepsilon(t)) + \int_0^t (\frac{1}{2}|u_n^\varepsilon(s)|^2 + g(x_n^\varepsilon(s)))ds$$
$$- \int_0^t(\phi_{n+1}^\varepsilon(x_n^\varepsilon(s)) - \phi_n^\varepsilon(x_n^\varepsilon(s))ds \tag{9.42}$$
$$+ \int_0^t R_n^\varepsilon(x_n^\varepsilon(s))ds$$

where $u_n^\varepsilon = - (\phi_{n+1}^\varepsilon)_x(x_n^\varepsilon)$.

Next by (9.40) we see that

$$\frac{1}{2}\frac{d}{dt}|x_n^\varepsilon(t) - x_n^\lambda(t)|^2 + \phi_{n+1}^\varepsilon(x_n^\varepsilon(t)) - \phi_{n+1}^\lambda(x_n^\varepsilon(t))$$
$$+ \phi_{n+1}^\lambda(x_n^\lambda(t)) - \phi_{n+1}^\varepsilon(x_n^\lambda(t)) \leq 0 \quad \text{a.e. } t > 0.$$

Since for $\varepsilon \to 0$, $\phi_{n+1}^\varepsilon \to \phi_{n+1}$ in $C(H)$ and again by (9.40), $|x_n^\varepsilon(t)| \leq |x_0|$ the latter implies that $\lim_{\varepsilon\to 0} x_n^\varepsilon(t) = x_n(t)$ exists uniformly on $\mathbb{R}^+$. Thus letting ε tend to zero in (9.42) we see that

$$\phi_{n+1}(x_0) \geq \phi_{n+1}(x_n(t)) + \int_0^t (\frac{1}{2}|u_n(s)|^2$$
$$+ g(x_n(s)))ds - \int_0^t (\phi_{n+1}(x_n(s)) - \phi_n(x_n(s)))ds \quad \forall t \geq 0 \tag{9.43}$$

where u_n is the weak limit of u_n^ε in $L^2(\mathbb{R}^+;H)$. Clearly we have

$$x_n' = Ax_n + u_n \qquad \text{a.e.} \quad t > 0 \tag{9.44}$$
$$x_n(0) = x_0.$$

Now since $\{\phi_n\}$ is a monotonically increasing sequence of continuous functions which converges to a continuous function we may infer by Dini's theorem that $\{\phi_n\}$ is uniformly convergent on bounded subsets. Then by (9.43) we see that:

$$\phi(x_0) \geq \liminf_{n \to \infty} \int_0^\infty (\tfrac{1}{2}|u_n|^2 + g(x_n))ds \geq \phi^\infty(x_0). \tag{9.45}$$

Now let (x^*,u^*) be an optimal pair in Problem (9.5). Then as seen in the proof of Theorem 12, $u^*(t) = -\phi_x^\infty(x^*(t))$ for $t \geq 0$ and x^* is the solution to the system

$$x^{*\prime} = Ax^* - \phi_x^\infty(x^*) \qquad t \geq 0 \tag{9.46}$$
$$x^*(0) = x_0.$$

Multiplying (9.46) by $(\phi_{n+1}^\varepsilon)_x(x^*)$ and using Eq. (9.41) we get

$$\frac{d}{dt}\phi_{n+1}^\varepsilon(x^*(t)) = (Ax^*(t) - \phi_x^\infty(x^*(t)),(\phi_{n+1}^\varepsilon)_x(x^*(t))$$

$$= -(\phi_x^\infty(x^*(t)),(\phi_{n+1}^\varepsilon)_x(x^*(t)) + \phi_{n+1}^\varepsilon(x^*(t)) - \phi_n^\varepsilon(x^*(t))$$

$$+ \frac{1}{2}|(\phi_{n+1}^\varepsilon)_x(x^*(t))|^2 - g(x^*(t)) - R_n^\varepsilon(x^*(t)) \quad \text{a.e.} \quad t > 0$$

and therefore

$$\frac{d}{dt}\phi_{n+1}^\varepsilon(x^*(t)) \geq -g(x^*(t)) - \frac{1}{2}|u^*(t)|^2 + \phi_{n+1}^\varepsilon(x^*(t))$$

$$- \phi_n^\varepsilon(x^*(t)) - R_n^\varepsilon(x^*(t)) \quad \text{a.e.} \quad t > 0.$$

Finally,

$$\phi_{n+1}^\varepsilon(x_0) \leq \phi_{n+1}^\varepsilon(x^*(t)) + \phi^\infty(x_0) \quad \forall t \geq 0$$

because $\phi_{n+1}^\varepsilon \geq \phi_n^\varepsilon$. Since $x^*(t_n) \to 0$ at least on some sequence $t_n \to \infty$ and $\phi_{n+1}^\varepsilon \in C(H)$ we conclude that $\phi_{n+1}(x_0) \leq \phi^\infty(x_0)$. Hence $\phi \leq \phi^\infty$ on H which along with (9.45) implies that $\phi = \phi^{\infty}$ as claimed.

Proof of Theorem 14 (continued)

Let $\{e_i\}$ be an orthonormal basis of H such that $Ae_i = \lambda_i e_i$, i = 1,2,... For each natural number N consider the space E_N generated by $(e_1,\dots,e_N)$ and by P_N the projection operator

$$P_N \cdot x = \sum_{i=1}^{N} (x,e_i)e_i.$$

Let $\phi = \lim_{n\to\infty} \phi_n$ be the function defined by (9.39) and ϕ^{∞} the function given by formula (9.5). On the space E_N define the functions ϕ_N^{∞} and ϕ_N,

$$\phi_N^{\infty}(P_N x) = \phi^{\infty}(P_N \cdot x) \quad \forall x \in H$$

respectively

$$\phi_N(P_N x) = \phi(P_N x) \quad \forall x \in H.$$

Let $A_N : E_N \to E_N$ be the linear operator

$$A_N P_N x = P_N A P_N x = A P_N x \quad \forall x \in H$$

substituting x by $P_N x$ in Eq. (9.1) we see that

$$\frac{1}{2}|(\phi_N^{\infty})_x(x)|^2 - (A_N x,(\phi_N^{\infty})_x(x)) = g_N(x) \quad \forall x \in E_N \tag{9.47}$$

where $g_N(P_N x) = g(P_N x)$ $\forall x \in H$.

By Lemma 15 and its proof also ϕ_N satisfies Eq. (9.47) and therefore $\phi_N = \phi_N^{\infty}$ for every N. Hence $\phi = \phi^{\infty}$ as claimed.

REMARK 11. $1^{\underline{o}}$ Assumptions of Theorem 14 are in particular satisfied if A is a self-adjoint positive operator in H with the domain D(A) compactly embedded in H.

$2^{\underline{o}}$ Theorem 11 and its proof is related to some general results established in [4], [9].

We shall conclude this section with an existence result for the stationary Riccati equation:

$$A^*P + PA - PBN^{-1}B^*P + M = 0. \tag{9.48}$$

Here we assume (δ), (β), (γ) (see §8) and (ε) $M \geq 0$. Moreover, we denote by $P = \Pi(P_0,M)$ the mild solution to Problem (8.1) and by $P_n = \Pi_n(P_0,M)$ the solution to Problem (8.4). We state first a lemma.

LEMMA 16. *Let* $P_0,\bar{P}_0 \in \Sigma^+(H)$, $G\ \bar{G} \in C_s([0,T];\Sigma^+(H))$, $P_0 \leq \bar{P}_0$, $G(t) \leq G(t)$, *then* $\Pi(P_0,G) \leq \Pi(\bar{P}_0,\bar{G})$.

Proof Let $P = \Pi(P_0,G)$, $\bar{P} = \Pi(\bar{P}_0,\bar{G})$, $P_n = \Pi_n(P_0,G)$, $\bar{P}_n = \Pi_n(\bar{P}_0,\bar{\sigma}_0)$, $Z_n = P_n-\bar{P}_n$; we have:

$$\begin{cases} Z_n' - \{A_n^* - \dfrac{(P_n+\bar{P}_n)}{2} BN^{-1}B^*\}Z_n - Z_n\left\{A_n - BN^{-1}B^* \dfrac{(P_n+\bar{P}_n)}{2}\right\} = G - \bar{G} \\ Z_n(0) = \bar{P}_0 - P_0 \end{cases}$$

it follows (consider the evolution operator associated with

$$\left\{A_n - BN^{-1}B^* \frac{(P_n+\bar{P}_n)}{2}\right\}_{t \in [0,T]})$$

that $Z_n(t) \geq 0$ and the claim follows from Theorem 11.

THEOREM 15. *Assume that* (β), (δ), (γ), (ε) *hold and moreover that there exists* $Q_o \in \Sigma^+(H)$ *such that,* $Q_o \in D(\mathbb{A})$ *(the operator defined in* (8.12)) *and*

$$\mathbb{A}(Q_0) - Q_0BN^{-1}B^*Q_0 + M \leq 0. \tag{9.49}$$

Then there exists the limit

$$\lim_{t\to\infty} \Pi(Q_0,M)(t)x = \bar{P}x. \tag{9.50}$$

Proof Set $Q(t) = Q_0$ then

$$Q' - \mathbb{A}(Q) + QBN^{-1}B^*Q \geq 0.$$

Moreover let $P = \Pi(Q_0,M)$; in virtue of Lemma 16 we have $P(t) \leq Q_0$. If $P_s(t) = P(t+s)$, that is $P_s = \Pi(P(s),M)$ we have (again by the lemma) $P_s(t) \leq P(t)$ consequently $P(t)$ is decreasing and the limit (9.50) exists.

Now $P_n(t) = P(t+n)$ is a mild solution to

$$P_n' - \mathbb{A}(P_n) = M - P_n^2$$

$$P_n(0) = P(n).$$

But $P_n(0) \to \bar{P}$, $M - P_n^2 \to M - \bar{P}^2$ in $C_s([0,T];\Sigma^+(H))$, it follows $P_n(t)x \to R(t)x = \bar{P}x$, where R is the mild solution to $R' - \mathbb{A}(R) = M - \bar{P}^2$; so $R = \bar{P}$ fulfils (9.49).

REMARK 12. 1º If B = 1 (9.49) is fulfilled with $Q_0 = \lambda I$ for a suitable λ.

2º Arguing as for Theorem 12 we see that if (A,B) is stabilizable (for example if B = 1) then, (under the hypotheses of Theorem 15) Eq. (9.48) has a unique solution.

3º The proof of Theorem 15 uses some arguments due to Tartar ([25]).

4º There exists an extensive literature on the stationary Riccati equation and its connections with control problems with infinite horizon. We cite in this context [1] and the works of D.L. Luxes and D.L. Russel [23], R. Datko [18], J. Zabczyk [27] (see the book [21] of Lions for further references and information on the subject).

§10 References

[1] M. De Acutis, Asymptotic behaviour for a problem arising in the optimal control theory. *J. Math. Anal. Appl. 77* 1980, 635-641.

[2] S. Aizawa, A semigroup treatment of the Hamilton-Jacobi equation in one space variable. *Hiroshima Math. J. 3*, 1973, 367-386.

[3] V. Barbu, *Nonlinear semigroups and differential equations in Banach spaces*, Noordhoff, 1976.

[4] V. Barbu, Convex control problems and Hamiltonian systems on infinite intervals. *SIAM J. Control and Optimiz 16*, 1978, 895-911.

[5] V. Barbu, Th. Precupanu, *Convexity and optimization in Banach spaces*, Sijthoff Noordhoff, Amsterdam, 1978.

[6] V. Barbu, G. Da Prato, Global existence for nonlinear operator equation arising in synthesis of optimal control, *Nonlinear Analysis 4*, 1980, 1157-1166.

[7] V. Barbu, G. Da Prato, Global existence for the Hamilton-Jacobi equations in Hilbert spaces, *Annali Scuola Normale Superiore Pisa*, VIII, 1981, 257-284.

[8] V. Barbu, G. Da Prato, Local existence for a nonlinear operator equation arising in synthesis of optimal control, *Numer. Functional Analysis and Optimization 1*, 1979, 665-677.

[9] V. Barbu, G. Da Prato, Existence and approximation for stationary Hamilton-Jacobi equations, *Nonlinear Analysis 5*, 1981, 1213-1224.

[10] B.C. Burch, A semigroup treatment of the Hamilton-Jacobi equations in several space variables, *J. Differential Equations 23*, 1977, 102-124.

[11] B.C. Burch, J. Goldstein, Some boundary value problems for the Hamilton-Jacobi equations, *Hiroshima Math. J. 8*, 1978, 223-233.

[12] S. Benton, *The Hamilton-Jacobi equation, a global approach*, Academic Press, New York, 1977.

[13] E. Conway, E. Hopf, Hamilton's theory and generalized solutions to the Hamilton-Jacobi equations. *J. Math. Mech.* 13, 1964, 936-986.

[14] M.G. Crandall, The semigroup approach to first order quasilinear equations in several space variables, *Israel J. Math. 12*, 1972, 108-132.

[15] M.G. Crandall, P.L. Lions, Condition d'unicité pour les solutions généralisées des équations de Hamilton-Jacobi du 1er ordre. *Comptes Rendus*, Paris, 252, 1981, 183-186.

[16] G. Da Prato, Quelques résultats d'existence, unicité et regularité pour un problème de la théorie du contrôle. *J. Math. Pures at Appl. 52*, 1973, 353-375.

[17] G. Da Prato, Some remarks on an operational time dependent equation. *Rend. Sem. Mat. Padova 59*, 247-262, 1978.

[18] R. Datko, Unconstrained control problems with quadratic cost. *SIAM J. Control 11*, 1973, 32-52.

[19] A. Douglis, Solutions in the large for multi-dimensional non linear partial differential equations of the first order. *Ann. Inst. Fourier 15*, 1965, 1-35.

[20] W.H. Fleming, The Cauchy problem for a nonlinear first order partial differential equation. *J. Differential Equations*, 1969, 515-530.

[21] J.L. Lions, *Optimal control of systems governed by partial differential equations*, Springer, Wiesbaden, 1972.

[22] P.L. Lions, *Generalized solutions of Hamilton-Jacobi equations*, Pitman, London, 1982.

[23] D.L. Lukes, D.L. Russel, The quadratic criterion for distributed systems. *SIAM J. Control 7*, 1969, 101-121.

[24] S.N. Kružkov, The Cauchy problem in the large for nonlinear equations and for certain quasi-linear systems of first order in several variables. *Soviet Math. Dokl 5*, 1964, 493-496.

[25] L. Tartar, Sur l'étude direct d'équations non linéaires intérvenant en théorie du contrôle optimal. *J. Functional Analysis 17*, 1974, 1-47.

[26] R. Teman, Sur l'équation de Riccati associée à des operateurs non bornés en dimension infinie. *J. Functional Analysis 7*, 1971, 85-116.

[27] J. Zabczyk, Remarks on the algebraic Riccati equations in Hilbert space. *Appl. Math. and Optimization 2*, 1976, 251-258.

3 Existence theory in nonconvex classes

§1 Introduction

In this chapter we propose studying existence for the Hamilton-Jacobi equations of the form:

$$\phi_t(t,x) + F(\phi_x(t,x)) - (Ax + \Gamma x, \phi_x(t,x))$$

$$= g(t,x);\ t \in [0,T],\ x \in D(A) \qquad (1.1)$$

$$\phi(0,x) = \phi_0(x)$$

in a real Hilbert space H endowed with the norm $|.|$ and the scalar product $(\cdot,\cdot)$. Here A is the infinitesimal generator of a C_0-semigroup of linear continuous operators on H and $\phi_0 : H \to \mathbb{R}$, $g:[0,T] \times H \to \mathbb{R}$, $F:H \to \mathbb{R}$, $\Gamma:H \to H$ are given mappings.

A direct approach to this problem developed in §2 gives the existence and uniqueness of a local smooth solution ϕ for ϕ_0, Γ, g sufficiently smooth and $F(y) = \frac{1}{2}|Dy|^2$ where $D \in \mathcal{L}(H,H)$.

As in the case of initial value problem for the finite dimensional nonlinear hyperbolic systems, Problem (1.1) cannot be solved generally in the class of smooth functions on large intervals in time. Even if ϕ_0, Γ, g are smooth, Eq. (1.1) develops singularities in finite time. The variational method developed in §4 gives the existence of global generalized solutions for Problem (1.1). Roughly speaking a generalized solution to Problem (1.1) is a locally Lipschitz function ϕ which satisfies (1.1) a.e. in t and for all x in D(A), ϕ_x being replaced by the generalized gradient in the sense of Clarke of $x \to \phi(t,x)$.

The application includes the existence of optimal feedback controllers for distributed control problems (§3, §6) and existence for stationary Hamilton-Jacobi equations (§5). The material presented in this chapter is closely related to the contents of the works [4], [5], [6], [7], [8].

§2 A direct approach to the Hamilton-Jacobi equations in a nonconvex class of solutions

We shall study here the initial value problem:

$$\phi_t(t,x) + \frac{1}{2}|D\phi_x(t,x)|^2 - (Ax + \Gamma x, \phi_x(t,x)) = g(t,x)$$

$$\text{for } x \in D(A) \text{ and } t \in [0,T] \qquad (2.1)$$

$$\phi(0,x) = \phi_0(x)$$

under the following assumptions:

$$g \in B([0,T];C^2_{Lip,\infty}(H)) \qquad (2.2)$$

$$\phi_0 \in C^2_{Lip,\infty}(H) \qquad (2.3)$$

$$\Gamma \in C^2_{Lip,\infty}(H,H);\ D \in \mathcal{L}(H,H). \qquad (2.4)$$

As noticed in the Introduction, A is the infinitesimal generator of a C_0-semigroup on H.

By a solution to Problem (2.1) we mean a continuous function $\phi[0,T] \times H \to \mathbb{R}$ which is differentiable in t for every $x \in D(A)$, Fréchet differentiable in x for every $t \in [0,T]$ and satisfies (2.1) for all $t \in [0,T]$ and $x \in D(A)$.

THEOREM 1. *Let* A, D, ϕ_0, Γ *and* g *satisfy the above assumptions. Then for* T *sufficiently small, Problem (2.1) has a unique solution* $\phi \in B([0,T];C^1_{Lip,\infty}(H))$.

In applications the following variant of Theorem 1 is useful.

THEOREM 2. *Assume that:*

$$g \in B([0,T];C^2_{Lip}(H)) \qquad (2.5)$$

$$\phi_0 \in C^2_{Lip}(H),\ \Gamma \in C^2_{Lip}(H,H) \qquad (2.6)$$

and A,D *are as in Theorem 1. Then for every* $r > 0$ *there exist* $T_r \in]0,T[$ *and* $\phi_r \in B([0,T_r];C^1_{Lip,\infty}(H))$ *which is the unique solution to (2.1) on* $[0,T_r] \times \Sigma_r$.

Theorem 2 follows from Theorem 1 by the following argument. For a fixed r we choose an infinitely differentiable function $\alpha_r:\mathbb{R} \to \mathbb{R}$ such that $\alpha_r(u) = 1$

for $|u| \leq r^2$ and $\alpha_r(u) = 0$ for $|u| \geq r^2 + 1$.

Now we apply Theorem 1 where g, ϕ_0 and Γ are taken as

$$g_r(t,x) = g(t,x)\alpha_r(|x|^2),\ \phi_{0,r}(x) = \phi_0(x)\alpha_r(|x|^2)$$

$$\Gamma_r(x) = \alpha_r(|x|^2)\Gamma(x) \text{ for } x \in H,\ t \in [0,T].$$

Then the corresponding solution ϕ_r to (2.1) which exists on some interval $[0,T_r]$ will satisfy Eq. (2.1) on $[0,T_r] \times \Sigma_r$ where $\Sigma_r = \{x \in H;\ |x| \leq r\}$.

In order to prove Theorem 1 we shall use a method similar to that employed in Chapter 2. The idea is to approximate Eq. (2.1) by

$$\phi_t(t,x) + \varepsilon^{-1}(\phi(t,x) - \phi_\varepsilon(t,x)) - (Ax + \Gamma x, \phi_x(t,x))$$

$$= g(t,x);\ x \in D(A),\ t \in [0,T] \tag{2.7}$$

$$\phi(0,x) = \phi_0(x), \text{ for } x \in H$$

where for $\phi \in C^1_{\mathrm{Lip},\infty}(H)$ and $x \in H$,

$$\phi_\varepsilon(x) = \phi(x_\varepsilon) + \varepsilon/2\ |D\phi^1(x_\varepsilon)|^2 \tag{2.8}$$

$$x_\varepsilon = (1 + \varepsilon D^*D\phi')^{-1}(x),\quad 0 < \varepsilon < |D|^{-2}|\phi|_1^{-1}. \tag{2.9}$$

We preface the proof of Theorem 1 with some simple properties of ϕ_ε and x_ε defined above.

Notice first that for $\phi \in C^2_{\mathrm{Lip},\infty}(H)$ and

$$0 < \varepsilon < |D|^{-2}|\phi|_2^{-1} \tag{2.10}$$

the map $x \to x_\varepsilon$ is differentiable on H and its differential $x'_\varepsilon(x) \in \mathcal{L}(H,H)$ is given by

$$x'_\varepsilon(x) = (1 + \varepsilon D^*D\phi''(x_\varepsilon))^{-1}. \tag{2.11}$$

LEMMA 1. *Let* $\phi, \tilde{\phi} \in C^2_{\mathrm{Lip},\infty}(H)$ *be given and let*

$$0 < \varepsilon < \inf\{|D|^{-2}|\phi|_2^{-1}; |D|^{-2}|\tilde{\phi}|_2^{-1}\}$$

Then

$$|x_\varepsilon - y_\varepsilon|_0 \leq (1 - \varepsilon|D|^2|\phi|_2)^{-1}|x - y| \tag{2.12}$$

$$|x_\varepsilon - \tilde{x}_\varepsilon|_0 \leq \varepsilon|D|^2(1 - \varepsilon|D|^{-2}|\phi|_2)^{-1}|\phi-\tilde{\phi}|_1 \tag{2.13}$$

$$|x_\varepsilon|_1 \leq (1 - \varepsilon|D|^2|\phi|_2)^{-1} \tag{2.14}$$

$$\|x_\varepsilon\|_1 \leq \varepsilon|D|^2 \|\phi\|_2 (1 - \varepsilon|D|^2|\phi|_2)^{-3} \tag{2.15}$$

$$|x_\varepsilon - \tilde{x}_\varepsilon|_1 \leq \varepsilon^2|D|^4\|\tilde{\phi}\|_2 (1 - \varepsilon|D|^2|\phi|_2)^{-2} \tag{2.16}$$

$$(1 - \varepsilon|D|^2|\tilde{\phi}|_2)^{-1}|\phi - \tilde{\phi}|_1 + \varepsilon|D|^2(1 - \varepsilon|D|^2|\phi|_2)^{-1}$$

$$(1 - \varepsilon|D|^2|\tilde{\phi}|_2)^{-1}|\phi - \tilde{\phi}|_2.$$

The proof is an immediate application of Eq. (2.11) and it is left to the reader.

Now differentiating Eq. (2.8) we see that

$$\phi_\varepsilon'(x) = \phi'(x_\varepsilon) \text{ for all } x \text{ in } H \tag{2.17}$$

$$\phi_\varepsilon''(x) = \phi''(x_\varepsilon)x_\varepsilon' \tag{2.18}$$

LEMMA 2. *Let* $\phi, \tilde{\phi} \in C^2_{Lip,\infty}(H)$ *and* $0 < \varepsilon < \inf\{|D|^{-2}|\phi|_2^{-1}, |D|^{-2}|\tilde{\phi}|_2^{-1}\}$.

Then

$$|\phi_\varepsilon|_0 \leq |\phi|_0 \tag{2.19}$$

$$|\phi_\varepsilon - \tilde{\phi}_\varepsilon|_0 \leq |\phi - \tilde{\phi}|_0 \tag{2.20}$$

$$|\phi_\varepsilon|_1 \leq |\phi|_1; \ |\phi_\varepsilon|_2 \leq (1 - \varepsilon|D|^2|\phi|_2)^{-1}|\phi|_2 \tag{2.21}$$

$$|\phi_\varepsilon - \tilde{\phi}_\varepsilon|_1 \leq (1 - \varepsilon|D|^2|\phi|_2)^{-1}|\phi - \tilde{\phi}|_1 \tag{2.22}$$

$$|\phi_\varepsilon|_2 \leq (1 - \varepsilon|D|^2|\phi|_2)^{-1}|\phi|_2 \tag{2.23}$$

$$\|\phi_\varepsilon\|_2 \leq (1 - \varepsilon|D|^2|\phi|_2)^{-3} \tag{2.23'}$$

$$|\phi_\varepsilon - \tilde{\phi}_\varepsilon|_2 \leq (1 - \varepsilon|D|^2|\phi|_2)^{-1} \tag{2.24}$$

$$(1 - \varepsilon|D|^2|\tilde{\phi}|_2)^{-1}(\varepsilon|D|^2\|\phi\|_2(1 - \varepsilon|D|^2|\phi|_2)^{-1}|\phi - \tilde{\phi}|_1$$

$$+ |\phi - \tilde{\phi}|_2).$$

Proof Inequalities (2.21)-(2.24) are immediate consequences of Lemma 1 and of Eqs (2.17), (2.18). To prove (2.19) and (2.20) we define for ϕ and ε, $\lambda > 0$ the function

$$\phi_{\varepsilon,\lambda}(x) = \inf\{|(\lambda + D^*D)^{-1/2}(x-y)|^2(2\varepsilon)^{-1} + \phi(y);\ y \in H\}. \tag{2.25}$$

By the Ekeland variational principle ([11]) there exists $\{x^n_{\varepsilon,\lambda}\} \subset H$ such that for $n \to \infty$,

$$\varepsilon^{-1}(\lambda + D^*D)^{-1}(x^n_{\varepsilon,\lambda} - x) + \phi'(x^n_{\varepsilon,\lambda}) \to 0 \tag{2.26}$$

$$(2\varepsilon)^{-1}|(\lambda + D^*D)^{-1/2}(x - x^n_{\varepsilon,\lambda})|^2 + \phi(x^n_{\varepsilon,\lambda}) \to \phi_{\varepsilon,\lambda}(x). \tag{2.27}$$

By (2.26) it follows that there exists $\{\theta_n\} \to 0$ such that

$$x^n_{\varepsilon,\lambda} + \varepsilon(\lambda + D^*D)\phi'(x^n_{\varepsilon,\lambda}) = x + \varepsilon(\lambda + D^*D)\theta_n.$$

Hence

$$x^n_{\varepsilon,\lambda} \to (1 + \varepsilon(\lambda + D^*D)\phi')^{-1}x = x_{\varepsilon,\lambda} \quad \text{for } n \to \infty$$

and by (2.27) it follows that

$$\phi_{\varepsilon,\lambda}(x) = \phi(x_{\varepsilon,\lambda}) + \varepsilon/2|(\lambda + D^*D)^{1/2}\phi'(x_{\varepsilon,\lambda})|^2.$$

By (2.25) we have

$$\inf\{\phi(y); y \in H\} \leq \phi_{\varepsilon,\lambda}(x) \leq \phi(x)$$

and therefore

$$\phi_{\varepsilon,\lambda}(x) - \tilde{\phi}_{\varepsilon,\lambda}(x) \leq \phi(\tilde{x}_{\varepsilon,\lambda}) - \tilde{\phi}(\tilde{x}_{\varepsilon,\lambda})$$

$$|\phi_{\varepsilon,\lambda}|_0 \leq |\phi|_0, \ |\phi_{\varepsilon,\lambda} - \tilde{\phi}_{\varepsilon,\lambda}|_0 \leq |\phi - \tilde{\phi}|_0.$$

Letting λ tend to zero in the latter we get (2.19) and (2.20) as claimed.

LEMMA 3. *Let* $\phi \in C^2_{Lip,\infty}(H)$ *be given. Then*

$$|\varepsilon^{-1}(\phi(x) - \phi_\varepsilon(x)) - \tfrac{1}{2}|D\phi'(x)|^2| \leq \varepsilon|D|^4|\phi|_1^2|\phi|_2 \tag{2.28}$$

$$|\varepsilon^{-1}(\phi'(x) - \phi'_\varepsilon(x) - \phi''(x)D^*D\phi'(x)|$$

$$\leq \varepsilon|D|^4|\phi|_1(|\phi|_2^2 + \|\phi\|_2|\phi|_1) \text{ for all } x \in H. \tag{2.29}$$

Proof By (2.28) we have

$$\varepsilon^{-1}(\phi(x) - \phi_\varepsilon(x)) - \tfrac{1}{2}|D\phi'(x)|^2 = \varepsilon^{-1}\int_0^1 (\phi'(\lambda x + (1 - \lambda)x_\varepsilon), x - x_\varepsilon)d\lambda$$

$$- \tfrac{1}{2}|D\phi'(x_\varepsilon)|^2 - \tfrac{1}{2}|D\phi'(x)|^2 = \int_0^1 (D\phi'(\lambda x + (1 - \lambda)x_\varepsilon), D\phi'(x_\varepsilon))d\lambda - \tfrac{1}{2}|D\phi'(x_\varepsilon)|^2 - \tfrac{1}{2}|D\phi'(x)|^2$$

$$= \frac{1}{2}\int_0^1 (D\phi'(\lambda x + (1 - \lambda)x_\varepsilon) - D\phi'(x_\varepsilon), D\phi'(x_\varepsilon))d\lambda$$

$$+ \frac{1}{2}\int_0^1 (D\phi'(\lambda x + (1 - \lambda)x_\varepsilon), D\phi'(x_\varepsilon) - D\phi'(x)d\lambda$$

$$+ \frac{1}{2}\int_0^1 (D\phi'(\lambda x) + (1 - \lambda)x_\varepsilon) - D\phi'(x), D\phi'(x)d\lambda.$$

Along with estimates (2.21), (2.22) the latter yields (2.28). As regards (2.29) we see that

$$\varepsilon^{-1}(\phi'(x) - \phi'_\varepsilon(x)) - \phi''(x)D^*D\phi'(x)$$

$$= \int_0^1 (\phi''(\lambda x + (1 - \lambda)x_\varepsilon) - \phi''(x))D^*D\phi'(x_\varepsilon)d\lambda$$

$$+ \phi''(x)D^*D(\phi'(x_\varepsilon) - \phi'(x))$$

and the desired inequality follows by (2.23) and (2.24).

Existence for approximating equation (2.7)

It is more convenient to work with an integral form of Eq. (2.7). With this in mind, we notice that a classical solution ϕ to the linear equation

$$\begin{aligned} &\phi_t(t,x) - (Ax + \Gamma x, \phi_x(t,x)) = h(t,x), \ t \in [0,T] \times \in D(A) \\ &\phi(0,x) = \phi_0(x) \end{aligned} \tag{2.30}$$

can be written in the form:

$$\begin{aligned} \phi(t,x) &= \phi_0(\exp(t(A + \Gamma))x) \\ &+ \int_0^t h(s,\exp((t - s)\ (A + \Gamma))x)ds \end{aligned} \tag{2.31}$$

where $z(t,x) = \exp(t(A + \Gamma))x)$ is the solution to the initial value problem

$$\begin{aligned} &z'(t) = (A + \Gamma)z(t), \quad 0 \leqslant t \leqslant \infty \\ &z(0) = x. \end{aligned} \tag{2.32}$$

In other words, $\exp(t(A + \Gamma))$ is the nonlinear semigroup generated by $A + \Gamma$. The standard existence results for nonlinear equations of the form (2.32) assure the existence and the differentiability of z. Moreover, we have

$$\begin{aligned} &|z_x(t,x)| \leqslant M_0, \ 0 \leqslant t \leqslant T, \quad x \in H \\ &|z_{xx}(t,x)| \leqslant M_0^3\ |\Gamma|_2 t, \ 0 \leqslant t \leqslant T, \quad x \in H \\ &\|z(t,\cdot)\|_1 \leqslant 3M_0^5\ |\Gamma|_2^2 t^2 + M_0^4\ \|\Gamma\|_2\, t;\ 0 \leqslant t \leqslant T, \quad x \in H \end{aligned} \tag{2.33}$$

where M_0 is a positive constant which depends on Γ and T.

For the sake of simplicity and without loss of generality we shall assume from now on that $M_0 = 1$.

LEMMA 4. *If* $h \in B([0,T];C_\infty^1(H))$ *then the function* ϕ *defined by Eq. (2.31) belongs to* $B([0,T];C_\infty^1(H))$ *and satisfies Eq. (2.30) on* $[0,T] \times D(A)$.

Conversely every solution $\phi \in B([0,T];C_\infty^1(H))$ *to Eq. (2.30) can be written in the form of Eq. (2.31).*

Proof By estimates (2.33) it is readily seen that $\phi \in B([0,T];C^1_\infty(H))$. To show that ϕ satisfies Eq. (2.30), assume first that A is continuous on H. Then $e^{t(A+\Gamma)}$ is a group of nonlinear operators on H and so $e^{-t(A+\Gamma)}$ makes sense. For every $x \in H$ and $t \in [0,T]$ by Eq. (2.31) we have

$$\phi(t,e^{-t(A+\Gamma)}x) = \phi_0(x) + \int_0^t h(s,e^{-s(A+\Gamma)}x)ds \tag{2.34}$$

and therefore

$$\phi_t(t,e^{-t(A+\Gamma)}x) - (\phi_x(t,e^{-t(A+\Gamma)}x),((A + \Gamma)e^{-t(A+\Gamma)}x))$$

$$= \frac{d}{dt}\phi(t,e^{-t(A+\Gamma)}x) = h(t,e^{-t(A+\Gamma)}x).$$

Substituting in the latter $e^{-t(A+\Gamma)}x$ by x we see that ϕ satisfies (2.30). Conversely if ϕ is a solution to Eq. (2.30) then the latter yields

$$\frac{d}{dt}\phi(t,e^{-t(A+\Gamma)}x) = h(t,e^{-t(A+\Gamma)}x).$$

Integrating over [0,t] we get (2.34) which is equivalent to Eq. (2.31).

In the general case, consider the Yosida approximation of $A, A_n = nA(nI-A)^{-1}$. It is well known that A_n is continuous, $A_nx \to Ax$ for $n \to \infty$ $x \in D(A)$ and

$$\exp(A_nt)x \to \exp(At)x \text{ uniformly on } [0,T] \text{ for } x \in H.$$

This yields

$$z_n(t,x) = \exp((A_n + \Gamma)t)x \to \exp((A + \Gamma)t)x$$

and

$$(z_n)_x(t,x) \to z_x(t,x) \text{ uniformly on } [0,T].$$

Let $\phi_n:[0,T] \times H \to \mathbb{R}$ be the function defined by Eq. (2.31) where $A = A_n$. We have for all $x \in H$

$$\phi_n(t,x) \to \phi(t,x)$$

$$(\phi_n)_x(t,x) \to \phi_x(t,x) \text{ uniformly on } [0,T]$$

while according to the first part of the proof

$$(\phi_n)_t(t,x) - A_n x + \Gamma x,(\phi_n)_x(t,x)) = h(t,x)$$

for all $t \in [0,T]$ and $x \in H$.

Thus letting n tend to $+\infty$ we see that ϕ satisfies Eq. (2.30) as claimed. Finally if ϕ is a solution to Eq. (2.30) we have

$$\phi_t(t,x) - (A_n x + \Gamma x,\phi_x(t,x)) = (Ax - A_n x,\phi_x(t,x)) + h(t,x)$$

and according to the first part of the proof

$$\phi(t,x) = \phi_0(\exp(t(A_n + \Gamma))x) + \int_0^t h(s,\exp((t-s)(A_n + \Gamma))x)ds$$
$$+ \int_0^t (Ax - A_n x,\phi_x(s,\exp((t-s)(A_n + \Gamma))x)ds$$

and for $n \to \infty$ we see that ϕ is given by Eq. (2.31). Thus the proof of Lemma 4 is complete.

In accordance with Lemma 4 we may represent Eq. (2.7) in the form

$$\phi(t,x) = e^{-t/\varepsilon}\phi_0(z(t,x)) + \int_0^t e^{-(t-s)/\varepsilon}(\varepsilon^{-1}\phi_\varepsilon + g)(s,z(t - s,x))ds. \tag{2.35}$$

LEMMA 5. *Under assumptions of Theorem 1 there exists* T *independent of* ε *and a unique function* $\phi^\varepsilon \in B([0,T];C^2_{Lip,\infty}(H))$ *which satisfies Eq. (2.35) on* $[0,T] \times H$.

Proof Let K be the closed subset of $B([0,T];C^2_\infty(H))$ defined by

$$K = \{\phi \in B([0,T];C^2_{Lip,\infty}(H)); |\phi(t,\cdot)|_{C^2_{Lip,\infty}(H)} \leq M \text{ for } t \in [0,T]\}$$

where M is a positive constant such that

$$|\phi|_{0\ C^2_{Lip,\infty}(H)} \leq M/2. \tag{2.36}$$

Let $\Gamma_\varepsilon : K \to B([0,T]; C^2_\infty(H))$ be defined by

$$(\Gamma_\varepsilon\phi)(t,x) = e^{-t/\varepsilon}\phi_0(z(t,x)) + \int_0^t e^{-(t-s)/\varepsilon}(\varepsilon^{-1}\phi_\varepsilon + g)(s,z(t-s,x))ds \tag{2.37}$$

we have

$$(\Gamma_\varepsilon\phi)_x(t,x) = e^{-t/\varepsilon}z^*_x(t,x)\phi_0'(z(t,x)) + \int_0^t e^{-(t-s)/\varepsilon}z^*_x(t-s,x)(\varepsilon^{-1}\phi_{\varepsilon,x} + g_x)(s,z(t-s,x))ds \tag{2.38}$$

$$(\Gamma_\varepsilon\phi)_{xx}(t,x) = e^{-t/\varepsilon}z^*_x(t,x)\phi_0''(z(t,x))z_x(t,x) + e^{-t/\varepsilon}z^*_{xx}(t,x)\phi_0'(z(t,x)) + \int_0^t e^{-(t-s)/\varepsilon}z^*_x(t-s,x) \tag{2.39}$$

$$(\varepsilon^{-1}\phi_{\varepsilon,xx} + g_{xx})(s,z(t-s,x))z_x(t-s,x)ds + \int_0^t e^{-(t-s)/\varepsilon}z^*_{xx}(t-s,x)(\varepsilon^{-1}\phi_{\varepsilon,x} + g_x)(s,z(t-s,x))ds.$$

Along with estimates (2.19), (2.21), (2.23) and (2.33) the latter yields

$$|(\Gamma_\varepsilon\phi)(t,\cdot)|_0 \leq e^{-t/\varepsilon}|\phi_0|_0 + \int_0^t e^{-(t-s)/\varepsilon}(\varepsilon^{-1}|\phi(s,\cdot)|_0 + |g(s,\cdot)|_0)ds \tag{2.40}$$

$$|(\Gamma_\varepsilon\phi)(t,\cdot)|_1 \leq e^{-t/\varepsilon}|\phi_0|_1 + \int_0^t e^{-(t-s)/\varepsilon}(\varepsilon^{-1}|\phi(s,\cdot)|_1 + |g(s,\cdot)|_1)ds \tag{2.41}$$

$$|(\Gamma_\varepsilon\phi)(t,\cdot)|_2 \leq e^{-t/\varepsilon}(|\phi_0|_2 + t|\Gamma|_2|\phi_0|_1) + \int_0^t e^{-(t-s)/\varepsilon}(\varepsilon^{-1}|\phi(s,\cdot)|_2)(1-\varepsilon|D|^2|\phi(s,\cdot)|_2)^{-1} \tag{2.42}$$

$$+ \int_0^t e^{-(t-s)/\varepsilon}(t-s)|\Gamma|_2(\varepsilon^{-1}|\phi(s,\cdot)|_1 + |g(s,\cdot)|_1)ds$$

$$\|(\Gamma_\varepsilon\phi)(t,\cdot)\|_2 \leq e^{-t/\varepsilon}(2t|\Gamma|_2|\phi_0|_2 + \|\phi_0\|_2 + 3t^2|\Gamma|_2^2$$

$$+ t \|\Gamma\|_2|\phi_0|_1 + t \|\Gamma\|_2|\phi_0|_2) + \int_0^t e^{-(t-s)/\varepsilon}((\varepsilon^{-1}(2(t-s)|\Gamma|_2|\phi(s,\cdot)|_2$$

$$(1 - \varepsilon|D|^2)|\phi(s,\cdot)|_2)^{-1} + (1 - \varepsilon|D|^2|\phi(s,\cdot)|_2)^{-3}\|\phi(s,\cdot)\|_2) \qquad (2.43)$$

$$+ 2(t - s)|\Gamma|_2|g(s,\cdot)|_2 + \|g(s,\cdot)\|_2)\,ds$$

$$+ \int_0^t e^{-(t-s)/\varepsilon}(\varepsilon^{-1}|\phi(s,\cdot)|_1 + |g(s,\cdot)|_1$$

$$+ (t - s)|\Gamma|_2(\varepsilon^{-1}|\phi(s,\cdot)|_2(1 - \varepsilon|D|^2|\phi(s,\cdot)|_2)^{-1}$$

$$+ g(s,\cdot)|_2))(3(t-s)|\Gamma|_2^2 + (t-s)\|\Gamma\|_2)\,ds.$$

By the above estimates we infer that for every $\varepsilon > 0$ there exists $t_\varepsilon \in]0,T]$ such that $\|(\Gamma_\varepsilon\phi)(t,\cdot)\|_{C^2_{\mathrm{Lip},\infty}(H)} \leq M$ for $t \in [0,t_\varepsilon]$ and all $\phi \in K$. Also it is easy to see that $\Gamma_\varepsilon\phi$, $(\Gamma_\varepsilon\phi)_x$ and $(\Gamma_\varepsilon\phi)_{xx}$ are continuous on $[0,T] \times H$. Hence Γ_ε maps K into itself. On the other hand, for $\phi^i \in K$, $i = 1,2$ we have by virtue of (2.20), (2.22) and (2.24)

$$\sup \{|(\Gamma_\varepsilon\phi^1)(t,\cdot) - (\Gamma_\varepsilon\phi^2)(t,\cdot)|_{C^2(H)}\ t \in [0,t_\varepsilon]\}$$

$$\leq C(1 - \exp(- t_\varepsilon/\varepsilon)) \sup \{|\phi^1(t,\cdot) - \phi^2(t,\cdot)|_2;\ t \in [0,t_\varepsilon]\}.$$

Hence for t_ε sufficiently small, Γ_ε is a contraction in the space $B([0,t_\varepsilon];C^2_\infty(H))$. Consequently Eq. (2.35) has a unique solution $\phi^\varepsilon \in K$.

By a standard argument it follows that there exists a maximal interval $[0,T_\varepsilon]$ on which Eq. (2.35) has a unique solution ϕ^ε in K. Estimates (2.40)-(2.43) imply via Gronwall's lemma the following equalities:

$$|\phi^\varepsilon(t,\cdot)|_0 \leq |\phi_0|_0 + t \sup \{|g(s,\cdot)|_0;\ s \in [0,T_\varepsilon]\}$$

$$|\phi^\varepsilon(t,\cdot)|_1 \leq |\phi_0|_1 + t \sup \{|g(s,\cdot)|_1;\ s \in [0,T_\varepsilon]\}$$

$$|\phi^\varepsilon(t,\cdot)|_2 \leq |\phi_0|_2 \exp((C_1 - 1)t/\varepsilon) + (C_1P_1\varepsilon)/C_1-1)$$
$$(\exp((C_1 - 1)t/\varepsilon) - 1)$$

$$\|\phi^\varepsilon(t,\cdot)\|_2 \leq \|\phi_0\|_2 \exp((C_2 - 1)t/\varepsilon) + (C_2P_2\varepsilon)/(C_2-1)$$
$$(\exp((C_2 - 1)t/\varepsilon) - 1)$$

where $C_1 = (1 - \varepsilon|D|^2M)^{-1}$, $C_2 = (1 - \varepsilon|D|^2M)^{-3}$ and P_i, $i = 1,2$ are independent of ε.

Thus there exists $T > 0$ independent of ε such that

$$\|\phi^\varepsilon(t,\cdot)\|_{C^2_{Lip,\infty}(H)} \leq 3M/4 \text{ for } 0 \leq t \leq \inf(T,T_\varepsilon).$$

The latter inequality implies that for all $\varepsilon > 0$, $T_\varepsilon > T$ (because T_ε is maximal with the property that $\phi^\varepsilon \in K$), thereby completing the proof of Lemma 5.

<u>The convergence of approximating solutions ϕ^ε</u>

We shall prove that for $\varepsilon \to 0$, $\{\phi^\varepsilon\}$ converges to a solution ϕ to Eq. (2.1). To this end we need the following lemma.

LEMMA 6. *Let* $\phi^{1,\varepsilon}$ *and* $\phi^{2,\varepsilon}$ *be two solutions to Eq. (2.35) corresponding to* ϕ_0^1, g^1 *and* ϕ_0^2, g^2 *respectively. Then on the common interval of existence* $[0,T]$ *one has*

$$|\phi^{1,\varepsilon}(t,\cdot) - \phi^{2,\varepsilon}(t,\cdot)|_{C^2(H)} \leq C(|\phi_0^1 - \phi_0^2|_{C^2_\infty(H)} \tag{2.44}$$
$$+ \sup\{|g^1(s,\cdot) - g^2(s,\cdot)|_{C^2_\infty(H)};\ s \in [0,T]\})\ t \in [0,T]$$

where C *is independent of* ε.

The proof is an immediate consequence of estimates (2.20), (2.22). To obtain (2.44) it suffices to use Gronwall's inequality in Eqs (2.37), (2.38) where $\Gamma_\varepsilon\phi$ is replaced by $\phi^{1,\varepsilon}$ and $\phi^{2,\varepsilon}$, respectively.

Let ε, $\lambda > 0$ be arbitrary and let ϕ^ε and ϕ^λ be the corresponding solutions to Eq. (2.35). We may regard ϕ^λ as the solution to Eq. (2.35) where g is replaced by

$$g + \lambda^{-1}(\phi^\lambda_\lambda - \phi^\lambda) - \varepsilon^{-1}(\phi^\lambda_\varepsilon - \phi^\lambda)$$

equivalently

$$g + (\lambda^{-1}(\phi^\lambda_\lambda - \phi^\lambda) - \tfrac{1}{2}|D\phi^\lambda_x|^2) + (\tfrac{1}{2}|D\phi^\lambda_x|^2 - \varepsilon^{-1}(\phi^\lambda_\varepsilon - \phi^\lambda)).$$

Then applying Lemmas 3 and 6 we obtain the estimate

$$|\phi^{\varepsilon}(t,\cdot) - \phi^{\lambda}(t,\cdot)|_{C^1_\infty(H)} \leqslant 3M^3|D|^4(\varepsilon + \lambda) \text{ for } t \in [0,T]. \tag{2.45}$$

Hence there exists a function $\phi \in B([0,T];C^1_{Lip,\infty}(H))$ such that

$$\lim_{\varepsilon\to 0} \phi^{\varepsilon}(t,\cdot) = \phi(t,\cdot) \text{ in } C^1_\infty(H) \text{ uniformly on } [0,T]. \tag{2.46}$$

It remains to show that ϕ satisfies Eq. (2.1). To this aim we notice that by Lemma 4 $\phi^{\varepsilon}(t,x)$ is differentiable in t for $x \in D(A)$ and

$$\phi^{\varepsilon}_t(t,x) + \varepsilon^{-1}(\phi^{\varepsilon}(t,x) - \phi^{\varepsilon}_{\varepsilon}(t,x)) - (Ax + \Gamma x,\phi^{\varepsilon}_x(t,x))$$

$$= g(t,x) \text{ for } t \in [0,T].$$

Along with inequality (2.28) the latter yields

$$\phi^{\varepsilon}_t + \frac{1}{2}|D\phi^{\varepsilon}_x|^2 - (Ax + \Gamma x,\phi^{\varepsilon}_x) - g \to 0 \text{ uniformly on } [0,T] \times H.$$

Then by (2.46) we may conclude that for every $x \in D(A)$, $\phi(t,x)$ is continuously differentiable in t on $[0,T]$ and satisfies Eq. (2.1).

For uniqueness we write Eq. (2.1) in the integral form

$$\phi(t,x) = e^{-t/\varepsilon}\phi_0(z(t,x)) + \int_0^t e^{-(t-s)/\varepsilon}(\varepsilon^{-1}\phi_\varepsilon + \varepsilon^{-1}(\phi - \phi_\varepsilon) - \frac{1}{2}|D\phi_x|^2 + g)(s,z)(t - s,x))ds. \tag{2.47}$$

If ϕ^1, ϕ^2 are two solutions to (2.1) (equivalently, to (2.47)) then estimate (2.28) and Gronwall's inequality yield

$$|\phi_1(t,x) - \phi_2(t,x)| \leqslant C\varepsilon|D|^4 \sup \{|\phi_1(s)|_1^2||\phi_1(s)||_1 + |\phi_2(s)|_1^2||\phi_2(s)||_1 ; s \in [0,T]\} \text{ for all } \varepsilon > 0.$$

Hence $\phi^1 = \phi^2$ as claimed. Thus the proof of Theorem 1 is complete.

REMARK 1. Theorem 1 has been established in [8]. A different approach to initial value Problem (2.1) has been given in [6] in the special case where D = 1 (the identity operator) and Γ is a potential operator, i.e., Γ^1 is self-adjoint. It must be observed that Eq. (2.1) for D = 1 is closely related to the quasilinear operator equation

$$\begin{aligned} &E_t(t,x) + E_x(t,x)(E(t,x) - Ax - \Gamma x) - A^*E(t,x) \\ &\qquad - \Gamma'(x)E(t,x) = g_x(t,x) \quad t \in [0,T],\ x \in D(A) \\ &E(0,x) = \phi_0'(x) \end{aligned} \tag{2.48}$$

which is obtained formally by differentiating Eq. (2.1) with respect to x. The underlying idea of the approach used in [6] is to prove the existence of a generalized solution E to Eq. (2.48) and to show that it is of the form $E = \phi_x$ where ϕ is a solution to Eq. (2.1). For existence in Eq. (2.48) one uses the approximating process given by the equation

$$\begin{aligned} &E_t - A^*E - E_xA + \varepsilon^{-1}(L - L_\varepsilon) = P \text{ on } [0,T] \times H \\ &E(0,x) = E_0(x) \end{aligned} \tag{2.49}$$

where $L = 1 - \Gamma$, $L_\varepsilon = L(1 + \varepsilon L)^{-1}$, $E_0 = \phi'$ and $P = g_x - \Gamma'\Gamma$. For other existence results on Eq. (2.48) we refer to [7].

On Riemannian manifolds a similar result has been given in [6].

§3 Applications to synthesis of optimal controls

We shall present here some consequence of Theorems 1 and 2 in existence of *optimal feedback controls* for control problems of the form: minimize

$$\int_0^T (g_0(t,x(t)) + \tfrac{1}{2}|u(t)|_U^2)dt + \phi_0(x(T)) \tag{3.1}$$

over all $u \in L^2(0,T;U)$ and $x \in C([0,T];H)$ subject to

$$\begin{aligned} &x' = Ax + \Gamma x + Bu, \quad t \in [0,T] \\ &x(0) = x \end{aligned} \tag{3.2}$$

$$|x(t)| \leq r \quad t \in [0,T] . \tag{3.3}$$

Here U is a real Hilbert space (the space of controls) with the norm denoted by $|\cdot|_U$ and scalar product $(\cdot,\cdot)_U$; r is any positive constant.

The functions g,ϕ_0 and the mapping $\Gamma:H \to H$, $A:H \to H$ satisfy the assumptions of Theorem 2 and $B \in \mathcal{L}(U,H)$. In other words,

$$g_0 \in B([0,T];C^2_{Lip}(H)),\ \phi_0 \in C^2_{Lip}(H),\ \Gamma \in C^2_{Lip}(H,H) \tag{3.4}$$

and A is the infinitesimal generator of a C_0-semigroup on H. Further we assume that for some constant ω,

$$((A + \Gamma)x,x) \leq \omega|x|^2 \quad \text{for all} \quad x \in D(A). \tag{3.5}$$

Our assumptions on A and Γ imply that the initial value problem (3.2) has a unique 'mild' solution $x \in C([0,T];H)$, i.e.,

$$x(t) = e^{tA}x_0 + \int_0^t e^{(t-s)A}(Bu(s) + \Gamma x(s))ds.$$

By (3.5) it follows that

$$|x(t)| \leq e^{\omega t}|x_0| + \int_0^t e^{\omega(t-s)}|Bu(s)|ds, \quad 0 \leq t \leq T. \tag{3.6}$$

We associate with Problem (3.1) the backward Hamilton-Jacobi equation (the dynamic programming equation associated with (3.1)).

$$\psi_t(t,x) - \tfrac{1}{2}|B^*\psi_x(t,x)|^2_U + (\psi_x(t,x),Ax + \Gamma x)$$

$$+ g_0(t,x) = 0;\ t \in [0,T],\ |x| \leq 2r \tag{3.7}$$

$$\psi(T,x) = \phi(x) \text{ for } |x| \leq 2r.$$

By the substitution $\phi(t,x) = \psi(T-t,x)$ Problem (3.7) reduces to (2.1) where $g(t,x) = g_0(T-t,x)$ and $D = (BB^*)^{1/2}$ (B^* is the adjoint of B). Thus by virtue of Theorem 2 there exists $T_r > 0$ such that for every $T \in]0,T_r[$ the initial value problem (3.7) has a unique solution ψ^T satisfying the condition

$$\psi^T \in B([0,T];C^1_{Lip,\infty}(H)). \tag{3.8}$$

Consider the closed loop differential equation

$$x' = Ax + \Gamma x - BB^*\,\psi^T_x(t,x),\ t \in [0,T]$$

$$x(0) = x_0 \tag{3.9}$$

By (3.4), (3.5) and (3.8) it follows that Eq. (3.9) has a unique 'mild' solution x^* which satisfies the estimate

$$|x^*(t)| \leq e^{\omega t}|x_0| + C\,|B|^2\omega^{-1}(e^{\omega t} - 1), \quad 0 \leq t \leq T \tag{3.10}$$

where $C = \sup\{|\psi^T(t,\cdot)|_1;\ t \in [0,T]\}$. Hence for $0 < T < T_r$ and

$$|x_0| < \delta(r) = (r + C|B|^2\omega^{-1})e^{-\omega T} - C|B|^2\omega^{-1} \tag{3.11}$$

the solution x^* remains in the closed ball Σ_r, i.e.,

$$|x^*(t)| \leq r \text{ for } t \in [0,T]$$

THEOREM 3. *Let* ψ^T *be the solution to (3.7) where* $0 < T < T_r$. *Let* x^* *be the solution to (3.9) where* $|x_0| \leq \delta(r)$. *Then* x^* *is an optimal state to Problem (3.1) corresponding to optimal control* $u = u^*$ *given by*

$$u^* = -B^*\psi_x(t,x^*(t)),\ 0 \leq t < T. \tag{3.12}$$

Moreover, one has

$$\psi^T(t,h) = \inf\{\int_t^T (g_0(s,x(s)) + \tfrac{1}{2}|u(s)|_U^2)ds + \phi_0(x(T))$$

$$x(t) = h,\ x' = Ax + \Gamma x + Bu,\ |x(s)| \leq r \tag{3.13}$$

$$\text{for } s \in [0,T] \ \text{ for } t \in [0,T],\ |h| \leq \delta(r).$$

Proof The proof is similar to that of Proposition 2 in Chapter 2. Let $u \in L^2(t,T;U)$ and let x be the corresponding 'mild' solution to Eq. (3.2) on $[t,T]$ with initial value condition $x(t) = h$ and such that $|x(s)| \leq r$ for $s \in [t,T]$. (By estimate (3.6) we see that for $|h| \leq \delta(r)$ such a control u always exists.) Let $\{h_\varepsilon\} \subset D(A)$ and $\{u_\varepsilon\} \subset C^1([t,T];U)$ be such that $h_\varepsilon \to h$ in H and $u_\varepsilon \to u$ in $L^2(t,T;H)$ for $\varepsilon \to 0$. Then for all ε sufficiently small, $|x_\varepsilon(s)| \leq 2r$ for $s \in [t,T]$. On the other hand, by (3.2) and (3.7) we have

$$\frac{d}{ds}\psi^T(s,x_\varepsilon(s)) = \psi_s^T(s,x_\varepsilon(s)) + (\psi_x^T(s,x_\varepsilon(s)),x'(s))$$

$$= -g_0(s,x_\varepsilon(s)) + \tfrac{1}{2}|B^*\psi_x^T(s,x_\varepsilon(s))|_U^2 - (Bu_\varepsilon(s),\psi_x^T(s,x_\varepsilon(s))$$

$$\geq - g_0(s,x_\varepsilon(s)) - \frac{1}{2}|u_\varepsilon(s)|_U^2 \text{ for } s \in [t,T].$$

Integrating the latter on [t,T] we get

$$\psi^T(t,h) \leq \int_t^T (g_0(s,x_\varepsilon(s)) + \frac{1}{2}|u_\varepsilon(s)|_U^2)ds + \phi_0(x_\varepsilon(T)).$$

Then letting ε tend to zero,

$$\psi^T(t,h) \leq \int_t^T (g_0(s,x(s)) + \frac{1}{2}|u(s)|_U^2 ds + \phi_0(x(T)). \tag{3.14}$$

Let us denote by S(t,h) the optimal value function of Problem (3.1), i.e., the function defined by the right-hand side of (3.13).

By (3.14) it follows that

$$\psi^T(t,h) \leq S(t,h) \text{ for all } t \in [0,T],\ |h| \leq \delta(r) < r. \tag{3.15}$$

Now we fix $h \in \Sigma_{\delta,r}$. Then by (3.10) and (3.11) we see that the solution $x = y(s)$ to the Cauchy problem

$$y' = Ay + \Gamma y - BB^*\psi_y^T(s,y),\ t < s \leq T$$

$$y(t) = h$$

remains in Σ_r. Arguing as above we may assume that y is differentiable. Then by (3.7) it follows that

$$\frac{d}{dt}\psi^T(s,y(s)) = g_0(s,y(s)) - \frac{1}{2}|B^*\psi_y^T(s,y)|_U^2, \text{ for } s \in [t,T].$$

Integrating over [t,T] we get

$$\psi^T(t,h) = \int_t^T (g_0(s,y(s)) + \frac{1}{2}|B^*\psi_y^T(s,y(s))|^2)ds + \phi_0(y(T))$$

and by (3.15) we may conclude that

$$\psi^T(t,h) = S(t,h) \text{ for all } t \in [0,T],\ |h| \leq \delta(r).$$

In particular for $t = 0$, $h = x_0$ we may infer by (3.13) that for $|x_0| \leq \delta(r) < r$ the solution x^* to (3.9) is an optimal state corresponding to optimal control u^* given by Eq. (3.12). Thus the proof of Theorem 3 is complete.

Theorem 3 amounts to saying that

$$u = - B^*\psi_y^T(t,x) \tag{3.16}$$

is an *optimal feedback control* for Problem (3.1). In particular we infer that *for* T *sufficiently small (3.1) has at least one optimal control* u^*.

By Theorem 3 it follows that if T is given and Γ, g, ϕ_0 are sufficiently small then the control Problem (3.1) admits an optimal feedback control on the form (3.16). More precisely, consider the control problem: minimize

$$\int_0^T (\lambda^2 g_0(x(t)) + \tfrac{1}{2}|u(t)|_U^2)dt + \lambda\phi_0(x(T)) \tag{3.17}$$

subject to

$$\begin{aligned} &x' = Ax + \lambda\Gamma x + Bu \qquad t \in [0,T] \\ &x(0) = x_0; \quad |x_0| < r \end{aligned} \tag{3.18}$$

$$|x(t)| < r, \ t \in [0,T] \tag{3.19}$$

where λ is a positive parameter and $g_0, \phi_0, A, \Gamma, B$ satisfy assumptions of Theorem 3. In addition we shall strengthen condition (3.5) to

$$(Ax + \Gamma x, x) < 0 \text{ for all } x \in D(A) \tag{3.20}$$

COROLLARY 1. *Let* $[0,T]$ *be a given interval. Then there exists* $\lambda_0 > 0$ *such that for* $\lambda \in]0,\lambda_0]$ *the conclusion of Theorem 3 remains valid for Problem (3.17).*

Proof We see that in Problem (3.1) Theorem 3 is applicable for $T = T_0$ where:

$$0 < T_0 < \inf \{T_r, (r - |x_0|)C^{-1}|B|^{-2}\}.$$

Changing the time scale by the substitution $t \to \lambda t$ we see that Problem (3.1) reduces to (3.17) where $T = \lambda^{-1}T_0$. We may conclude therefore that Theorem 3 is applicable on interval $[0,T]$ if

$$0 < \lambda < \lambda_0 = T^{-1} \inf \{T_r, (r - |x_0|)C^{-1}|B|^{-2}\}$$

as claimed.

We shall now present some specific examples.

Parabolic distributed control problems

Consider the following control problem: minimize

$$\frac{1}{2}\left(\int_Q h(t,x)|y(t,x) - y^0(t,x)|^2 dx\,dt + \int_0^T |u(t)|_U^2 dt\right) \tag{3.21}$$

over all $u \in L^2(0,T;U)$ and $y \in H^{2,1}(Q) \cap L^2(0,T;\Omega)$ subject to the nonlinear diffusion process described by the equation

$$\begin{aligned} &y_t - \Delta y + \beta(y) \ni Bu \text{ on } Q =]0,T[\times \Omega \\ &y(0,x) = y_0(x) \quad x \in \Omega \\ &y = 0 \text{ on } \Sigma =]0,T[\times \Gamma \end{aligned} \tag{3.22}$$

where Ω is an open and bounded subset of $\mathbb{R}^n$ having a sufficiently smooth boundary Γ, $B \in \mathcal{L}(U,L^2(\Omega))$ and $h \in L^\infty(Q)$, $y^0 \in L^2(Q)$, $y_0 \in H_0^1(\Omega)$ are given functions. β is a maximal monotone graph in $\mathbb{R} \times \mathbb{R}$ such that $\beta(y_0) \in L^2(\Omega)$. A large class of distributed control processes arising in heat conduction and temperature control can be written in this form. For instance, in the special case where β is defined by

$$\beta(r) = 0 \text{ for } r > 0,\ \beta(0) = R^-,\ \beta(r) = \emptyset \text{ for } r < 0.$$

Equation (3.22) reduces to a well-known free boundary value problem equivalent to the one-phase Stefen problem (the obstacle problem).

Of course Theorem 3 is not applicable in this form. However, Problems (3.21), (3.22) can be approximated by a control problem with cost functional (3.21) and state system

$$\begin{aligned} &y_t - \Delta y + \beta^\varepsilon(y) = Bu \text{ in } Q \\ &y(0,x) = y_0(x) \quad x \in \Omega \\ &y = 0 \text{ on } \Sigma \end{aligned}$$

where β^ε is a C^∞-approximation of β. For instance we may choose β^ε as

$$\beta^\varepsilon(r) = \int_{-\infty}^{+\infty} \beta_\varepsilon(r - \varepsilon\theta)\rho(\theta)d\theta \text{ for } r \in \mathbb{R} \tag{3.23}$$

where $\beta_\varepsilon = \varepsilon^{-1}(1 - (1 + \varepsilon\beta)^{-1})$ and ρ is a C^∞-mollifier on real axis. Problems (3.21) and (3.23) are of the form (3.1) where $H = L^2(\Omega)$, $Ay = -\Delta y, D(A) = H_0^1(\Omega) \cap H^2(\Omega)$, $(\Gamma y)(x) = \beta^\varepsilon(y(x))$ a.e. $x \in \Omega$; $y \in L^2(\Omega)$ and

$$g_0(t,y) = \int_\Omega h(t,x)|y(x) - y^0(t,x)|^2 dx;\ t \in [0,T],\ y \in L^2(\Omega).$$

<u>Control of nonlinear wave equation</u>

Consider the following optimization problem: minimize

$$\int_0^T (g_0(t,y(t),y'(t)) + \tfrac{1}{2}|u(t)|_U^2)dt + \phi_0(y(T),y'(T)) \tag{3.24}$$

subject to

$$\begin{aligned} &y''(t) + A_0y(t) + My(t) = B_0u(t), \quad t \in [0,T] \\ &y(0) = y_0,\ y'(0) = y_1 \end{aligned} \tag{3.25}$$

$$|y(t)|_V^2 + |y'(t)|_E^2 \leqslant r^2 \text{ for } t \in [0,T]. \tag{3.26}$$

Here V,E are Hilbert spaces such that $V \subset E \subset V'$ algebraically and topologically and V is a dense part of E (V' is the dual space of V).

By $|\cdot|_V$ and $|\cdot|$ we have denoted the norms of V and E, respectively and by $(\cdot,\cdot)_0$ we shall denote the pairing between V,V' and the scalar product of E. A_0 is a linear symmetric and continuous operator from V to V' such that for some $\omega > 0$

$$(A_0y,y) \geqslant \omega|y|_V^2 \text{ for all } y \in V$$

and $M:V \to E$ is a mapping of class C^2 on V with $M^{(k)}$, $k = 0,1,2$ bounded and Lipschitz on bounded subsets. B_0 is a linear continuous operator from a real Hilbert space U (the control space) to E and

$$g_0 \in B([0,T];\ C^2_{Lip}(V \times E)) \tag{3.27}$$

$$\phi_0 \in C^2_{Lip}(V \times E);\ y_0 \in V,\ y_1 \in E. \tag{3.28}$$

The product space $H = V \times E$ is endowed with the usual Hilbertian structure

$$\langle(v_1,v_1),(v_1,v_2)\rangle = (A_0v_1,v_2) + (y_1,y_2) \text{ for all } v_1,v_2 \in V,$$
$$y_1,y_2 \in E.$$

Setting $Y(t) = \{y(t),y'(t)\}$, $y_0 = \{y_0,y_1\}$ we may write Eq. (3.25) in the form:

$$Y'(t) = AY(t) + \Gamma Y(t) + Bu(t) \quad t \in [0,T]$$
$$Y(0) = Y_0 \tag{3.29}$$

where

$$A = \begin{Vmatrix} 0 & 1 \\ -A_0 & 0 \end{Vmatrix} \; ; \; \Gamma(y,z) = \{0, - My\} \text{ for all } (y,z) \in V \times E$$

$$Bu = \{0,B_0u\} \text{ for all } u \in U.$$

In terms of A,Γ and B the Hamilton-Jacobi equation (3.7) for Problems (3.24)-(3.26) becomes,

$$\psi_t(t,y,z) - \frac{1}{2}|B_0^*\psi_z(t,y,z)|_U^2 + (\psi_y(t,y,z),A_0z)$$
$$- (\psi_z(t,y,z),A_0y + My) + g_0(t,y,z) = 0 \text{ for all} \tag{3.30}$$
$$(y,z) \in D(A_0) \times V, \; t \in [0,T]$$
$$\psi(T,y,z) = \phi_0(y,z) \text{ for all } (y,z) \in V \times E$$

where $D(A_0) = \{y \in V; A_0y \in E\}$.

Let $\psi^T:[0,T] \times V \times E \to \mathbb{R}$ be the solution to Eq. (3.30). By Theorem 3 we may infer for T 'sufficiently small' (equivalently for g_0,ϕ_0, and M 'sufficiently small')

$$u = - B_0^*\psi_{y'}^T(t,y,y') \tag{3.31}$$

is an optimal feedback control for Problem (3.24).

REMARK 2. 1º In the special case $E = L^2(\Omega)$, $V = H_0^1(\Omega)$, $A_0 = - \Delta$ and $(My)(x) = y^3(x)$ a.e. $x \in \Omega$, (3.25) is the well-known Klein-Gordon equation

$$y_{tt} - \Delta y + y^3 = B_0 u \text{ on } Q$$

$$y(0,x) = y_0(x), y_t(0,x) = y_1(x), \quad x \in \Omega \tag{3.32}$$

$$y = 0 \text{ on } \Sigma =]0,T[\times \Gamma \ .$$

2º The equation

$$y_{tt} - \Delta y + \beta^{\varepsilon}(y) = B_0 u \text{ on } Q$$

$$y(0,x) = y_0(x), \ y_t(0,x) = y_1(x) \quad x \in \Omega \tag{3.33}$$

$$y = 0 \text{ on } \Sigma$$

where β^{ε} is defined by Eq. (3.23) and $\beta(r) = 0$ for $r > 0$, $\beta(0) = R^-$ represents a good approximation for the 'obstacle' problem associated with the wave equation,

$$y_{tt} - \Delta y = B_0 u \text{ on } \{y > 0\}, \ y_{tt} - \Delta y \geq B_0 u \text{ on } Q$$

$$y(0,x) = y_0(x), \ y_t(0,x) = y_1(x) \quad x \in \Omega \tag{3.34}$$

$$y = 0 \text{ on } \Sigma.$$

Problem (3.34) models the control of an elastic membrane $\Omega \subset \mathbb{R}^3$ fixed at Γ constrained to remain over the obstacle $\{y = 0\}$ and subject to a vertical field of forces with the density $B_0 u$.

Control of functional differential equations

We shall examine here the following control problem: minimize

$$\int_0^T (h(t,y(t)) + \frac{1}{2}|u(t)|_m^2)dt + \psi_0(y(T)) \tag{3.35}$$

over all $y:[-a,T] \to \mathbb{R}^n$ and $u \in L^2(0,T;\mathbb{R}^m)$ subject to

$$y'(t) = A_0 y(t-a) + f(y(t)) + B_0 u(t) \text{ a.e. } t \in]0,T[$$

$$y(0) = y_0, \ y(t) = y^0(t) \text{ a.e. } t \in]-a,0[\tag{3.36}$$

$$|y(t)|_n \leq r \quad \text{a.e. } t \in]-a,T[. \tag{3.37}$$

Here A_0 is a $n \times n$ constant matrix, B_0 is a $n \times m$ constant matrix,

$$h \in B([0,T];C^2_{Lip}(\mathbb{R}^n)),\ \psi_0 \in C^2_{Lip}(\mathbb{R}^n) \text{ and } f \in C^2_{Lip}(\mathbb{R}^n;\mathbb{R}^n).$$

By $|\cdot|_n$ we have denoted the Euclidean norm in $\mathbb{R}^n$. It is well known (see for instance [10]) that system (3.36) can be written as an evolution equation in an appropriate Hilbert space. To this end consider the product space

$$M^2(-a,0;\mathbb{R}^n) = \mathbb{R}^n \times L^2(-a,0;\mathbb{R}^n)$$

and the operators A and Γ defined on $M^2(-a,0;\mathbb{R}^n)$ by

$$A(y_0,y^0) = \{A_0y^0(-a),y^{0\prime}\} \text{ for } (y_0,y^0) \in D(A)$$

where

$$D(A) = \{(y_0,y^0) \in M^2(-a,0;\mathbb{R}^n);\ y^0 \in W^{1,2}(-a,0;\mathbb{R}^n);\ y^0(0) = y_0\}$$

and

$$\Gamma(y_0,y^0) = \{f(y_0),0\}.$$

Then setting $Y(t) = \{y(t),y_t\}$ where $y_t(s) = y(t+s)$ for $-a \leqslant s \leqslant 0$ and $Bu = \{B_0u,0\}$ we may write Eq. (3.36) as

$$\begin{aligned} &Y'(t) = AY(t) + \Gamma Y(t) + Bu(t) \quad t \in]0,T[\\ &Y(0) = \{y_0,y^0\} \end{aligned} \qquad (3.38)$$

Next we define $g_0:[0,T] \times M^2(-a,0;\mathbb{R}^n) \to \mathbb{R}$ and $\phi_0:M^2(-a,0;\mathbb{R}^n) \to \mathbb{R}$ by

$$g_0(t,Y) = h(t,y_0);\ \phi_0(Y) = \psi_0(y_0),\ Y = \{y_0,y^0\}.$$

The Hamilton-Jacobi equation (3.7) becomes:

$$\begin{aligned} &\psi_t(t,y,z) - \tfrac{1}{2}|B_0^*\psi_y(t,y,z)|_m^2 + (\psi_y(t,y,z),f(y) + A_0z(-a)) \\ &+ \int_{-a}^{0} (\psi_z(t,y,z),z'(s))ds + h(t,y) = 0 \quad t \in [0,T] \\ &\psi(T,y,z) = \psi_0(y),\quad |y|_n \leqslant r. \end{aligned} \qquad (3.39)$$

If ψ^T is the solution to Eq. (3.39) (which exists by Theorem 1) we may conclude by Theorem 3 that for T sufficiently small, the feedback law

$$u(t,y) = -B_0^*\psi_y^T(t,y,y_t)$$

is an optimal feedback control for Problem (3.35).

The synthesis of some bilinear control systems has been studied by a similar method in [9]. However, straightforward resolving of the *dynamic programming equation* (3.7) appears to be difficult even in simple cases. Numerical methods are needed in order to open the way of the above results toward applications. This aspect of the problem will be discussed in §6. Here we restrict ourselves to indicating a simple conceptual algorithm.

We shall assume that the functions ϕ_0,g and the mapping Γ are of class C^∞. Then we associate with Eq. (3.7) the following sequence of linear equations

$$\psi_t^i + (\psi_x^i, Ax + \Gamma x - BB^*\psi_x^{i-1}) + \frac{1}{2}|B^*\psi_x^{i-1}|_U^2 + g_0(t,x) = 0,$$

$$x \in H,\ t \in [0,T] \tag{3.40}$$

$$\psi^i(T,x) = \phi_0(x);\ x \in H$$

for i = 0,1,... . The solution ψ^i can be written as (see (2.31))

$$\psi^i(t,x) = \phi_0(\eta^i(T,t,x)) + \int_t^T g^i(s,\eta^i(s,t,x))ds \tag{3.41}$$

where

$$g^i(t,x) = g_0(t,x) + \frac{1}{2}|B^*\psi_x^{i-1}(t,x)|_U^2$$

and $\eta^i(s,t,x)$ is the solution to the initial value problem

$$y' = Ay + \Gamma y - BB^*\psi_x^{i-1}(s,y),\ t \leq s \leq T$$

$$y(t) = x.$$

A little calculation involving Eq. (3.40) reveals that

$$\psi_T(t,x) \leq \psi^i(t,x) \leq \psi^{i-1}(t,x) \text{ for all } (t,x) \in [0,T] \times H \tag{3.42}$$

where ψ_T is the value function of optimal control Problem (3.1) (see Eq. (3.13)). Hence

$$\psi^*(t,x) = \lim_{i\to\infty} \psi^i(t,x) \tag{3.43}$$

exists for all $(t,x) \in [0,T] \times H$.

By (3.40) we see that

$$\psi^i_t - \frac{1}{2}|B^*\psi^i_x|^2_U + (\psi^i_x, Ax + \Gamma x) + g_0(t,x) = \frac{1}{2}|B^*(\psi^i_x - \psi^{i-1}_x)|^2_U.$$

Thus if $\psi^i_x \to \psi^*_x$ for $i \to \infty$ we may infer that $\psi^* = \psi_T$ is just the solution to Eq. (3.7). In general this is not true, but we may nevertheless regard ψ^* as a generalized solution to Eq. (3.7) and

$$u = -B^*\psi^i_x(t,x) \tag{3.44}$$

as an approximating optimal feedback control for our problem. The feedback control (3.44) is suboptimal, however.

§4 A variational approach - Generalized solutions

We shall study here by variational methods the global existence for the initial value problem

$$\begin{aligned} &\phi_t(t,x) + \Phi(D\phi_y(t,y)) + (Ly,\phi_y(t,y)) = g(y) \\ &\qquad\qquad \text{for } t \in [0,T],\ y \in D(L) \\ &\phi(0,y) = 0 \end{aligned} \tag{4.1}$$

in the space $H = L^2(\Omega)$; $\Omega \subset \mathbb{R}^N$.

We have denoted by ϕ_t the derivative in t of the unknown function $\phi:[0,T] \times H \to \mathbb{R}$ and by ϕ_y the differential of ϕ in some generalized sense which will be defined later.

We shall assume that the following conditions are satisfied:

(i) D is a linear continuous operator from H to a real Hilbert space U.

(ii) $\Phi:U \to \mathbb{R}$ is a continuous convex function on U satisfying the growth condition

$$\Phi(u) \leq C(|u|^2_U + 1) \text{ for all } u \in U. \tag{4.2}$$

(iii) The function $g:H \to \mathbb{R}$ is locally Lipschitz and non negative.

(iv) $L:D(L) \subset H \to H$ is a nonlinear operator defined by

$$Ly = -Ay + \beta(y) \text{ for } y \in D(L) \tag{4.3}$$

$$D(L) = \{y \in D(A);\ \beta(y) \in H\} \tag{4.4}$$

where $\beta:\mathbb{R} \to \mathbb{R}$ is a locally Lipschitz monotonically increasing function which satisfies the conditions:

$$\beta'(r) \leq C(|\beta(r)| + |r| + 1) \text{ a.e. } r > 0,\ \beta(0) = 0 \tag{4.5}$$

and $A:D(A) \subset H \to H$ is a linear self-adjoint dissipative operator having the property that for every λ real the level set

$$\{y \in D(A);\ -(Ay,y) + |y|^2 \leq \lambda\} \tag{4.6}$$

is compact in H and

$$(Ay,\delta(y)) \leq C\ |\delta(y)|\ (1 + |y|) \quad \forall y \in D(A) \tag{4.7}$$

for every increasing and Lipschitz function $\delta:\mathbb{R} \to \mathbb{R}$ with the property that $\delta(0) = 0$.

The following two assumptions will be also used

(v) Either β is globally Lipschitz or $D(A) \subset H^2(\Omega)$ and $N = 1$.

(vi) Φ is Fréchet differentiable on H and $\nabla\Phi$ is locally Lipschitz.

As in the previous section we shall denote by $|\cdot|$ and $(\cdot,\cdot)$ the usual norm and scalar product in $H = L^2(\Omega)$. The norm of U will be denoted by $|\cdot|_U$ and the scalar product by $(\cdot,\cdot)_U$. It must be noticed that assumption (iv) implies that L is a maximal monotone operator in H and the domain D(L) is a dense subset of H (see Proposition 5 in Chapter 1). As a matter of fact L is the subdifferential ∂l of the lower semicontinuous convex function $l:H \to \bar{\mathbb{R}}$,

$$l(y) = \frac{1}{2}|(-A)^{1/2}y|^2 + \int_\Omega j(y(x))dx \text{ for all } y \in H \tag{4.8}$$

where j is the primitive of β.

A typical example is

$$Ay = \sum_{i,j=1}^{N} (a_{ij}(x)y_{x_i})_{x_j} \quad \text{for } y \in D(A) \tag{4.9}$$

$$D(A) = \{y \in H^2(\Omega);\ \alpha_1 y + \alpha_2 \frac{\partial y}{\partial \nu} = 0 \text{ on } \Gamma\}$$

where $a_{ij} \in C^1(\bar{\Omega})$, $a_{ij} = a_{ij}$ for all $i,j = 1,\dots,N$, $\alpha_1 + \alpha_2 > 0$ $\alpha_i \geqslant 0$ $i = 1,2$ and for some $\omega > 0$

$$\sum_{i,j=1}^{N} a_{ij}(x)\xi_i\xi_j \geqslant \omega|\xi|_N^2 \text{ for all } \xi \in \mathbb{R}^N,\ x \in \Omega.$$

We shall associate with Eq. (4.1) a control problem having (4.1) as a dynamic programming equation (see §7 in Chapter 2). This is the optimal control problem: minimize

$$(P) \int_0^T (g(y(s)) + \Phi^*(u(s)))ds + \psi(y(T))$$

over all $u \in L^2(0,T;U)$ and $y \in W^{1,2}(0,T;H) \cap L^2(0,T;D(A))$ subject to

$$\begin{aligned} &y'(t) + Ly(t) = Bu(t) \quad \text{a.e. } t \in]0,T[\\ &y(0) = y_0 \end{aligned} \tag{4.10}$$

where $y_0 \in D(1)$, i.e.,

$$y_0 \in D((-A)^{1/2});\quad j(y_0) \in L^1(\Omega). \tag{4.11}$$

Here $B = -D^*$,ψ is a locally Lipschitz function on H and Φ^* is the conjugate of Φ, i.e.,

$$\Phi^*(p) = \sup\{(p,u)_U - \Phi(u);u \in U\}. \tag{4.12}$$

It must be noticed that assumption (4.2) implies that for some $C_1 > 0$ and C_2 real

$$\Phi^*(u) \geqslant C_1|u|_U^2 + C_2 \text{ for all } u \in U. \tag{4.13}$$

Let us denote by $y(s,t,y_0,u)$ the solution to

$$\hat{y}'(s) + Ly(s) = Bu(s) \quad \text{a.e.} \quad s \in]t,T[$$

$$y(t) = y_0.$$

By Proposition 9 in Chapter 1 the mapping $(y_0,u) \to y(s,t,y_0,u)$ is well defined from $H \times L^2(t,T;U)$ to $C([t,T];H)$ and from $D(1) \times L^2(t,T;U)$ to $W^{1,2}(t,T;H) \cap L^2(t,T;D(L))$. Then by the Ascoli-Arzelà theorem the map $u \to y(\cdot,y_0,u)$ is compact from $L^2(0,T;U)$ to $C([0,T];H)$. Since the functional $u \to \int_0^T \Phi^*(u(s))ds$ is weakly lower semicontinuous on $L^2(0,T;H)$ (because it is convex and lower semicontinuous) the above mentioned property along with (4.13) implies that problem (P) has at least one optimal pair $(y^*,u^*) \in W^{1,2}(Q,T;H) \times L^2(0,T;U)$.

The main ingredient of our approach is a maximum principle type result for Problem (P).

PROPOSITION 1. *Let* (y^*,u^*) *be an optimal pair in Problem* (P). *Then there exists a function* $p \in AC([0,T];Y^*) \cap C_w([0,T];H) \cap L^\infty(0,T;H) \cap L^2(0,T;V)$ *such that*

$$p' + Ap \in L^1(Q) \tag{4.14}$$

$$p' + Ap \in p\,\partial\beta(y^*) + \partial g(y^*) \quad \text{a.e. in } Q \tag{4.15}$$

$$p(T) + \partial\psi(y^*(T)) \ni 0 \tag{4.16}$$

$$B^*p(t) \in \partial\Phi^*(u^*(t)) \quad \text{a.e.} \quad t \in]0,T[. \tag{4.17}$$

Here V is the space $D((-A)^{1/2})$ endowed with the Hilbertian norm $|y|_V^2 = |(-A)^{1/2}y|^2 + |y|^2$ and Y^* is the dual space of $Y = H^s(\Omega) \cap V$ where $s > N/2$. By $\partial\beta$, $\partial\psi$ and ∂g we have denoted the generalized gradients of β, ψ and g (see §2 in Chapter 1), $C_w([0,T];H)$ is the space of weakly continuous functions from [0,T] to H.

Proof of Proposition 1

The proof will be only outlined. For complete proof we refer the reader to [4], [5] (see also [3] for a general treatment of control problems governed by semilinear variational inequalities of parabolic type).

For any $\varepsilon > 0$ consider the approximating control problem: minimize

$$\int_0^T (g^\varepsilon(y(t)) + \Phi^*(u(t)) + \tfrac{1}{2}|u(t) - u^*(t)|_U^2)dt + \psi^\varepsilon(y(T)) \tag{4.18}$$

over all $u \in L^2(0,T;U)$ and $y \in W^{1,2}(0,T;H)$ subject to

$$y' - Ay + \beta^\varepsilon(y) = Bu \quad \text{a.e. } t \in]0,T[$$
$$y(0) = y_0 \tag{4.19}$$

where

$$\beta^\varepsilon(r) = \int_{-\infty}^{\infty} \beta_\varepsilon(r - \varepsilon\theta)\rho(\theta)d\theta; \ \beta_\varepsilon = \varepsilon^{-1}(1 - (1 + \varepsilon\beta)^{-1}) \tag{4.20}$$

and ρ is a C_0^∞-mollifier on $\mathbb{R}$. The functions g^ε and ψ^ε are defined as follows

$$g^\varepsilon(y) = \int_{\mathbb{R}^n} g(P_n y - \varepsilon\Lambda_n\tau)\rho_n(\tau)d\tau, \quad y \in H \tag{4.21}$$

$$\psi^\varepsilon(y) = \int_{\mathbb{R}^n} \psi(P_n y - \varepsilon\Lambda_n\tau)\rho_n(\tau)d\tau, \quad y \in H \tag{4.22}$$

where $P_n:H \to X_n$ is the projection operator on X_n, ρ_n is a C_0^∞-mollifier in $\mathbb{R}^n$ and $\Lambda_n:\mathbb{R}^n \to X_n$ is the operator

$$\Lambda_n(\tau) = \sum_{i=1}^{n} \tau_i e_i; \ \tau = (\tau_1,\ldots,\tau_n).$$

Here $\{e_i\}_{i=1}^\infty$ is an orthonormal basis in H and X_n is the space generated by $\{e_i\}_{i=1}^n$, $n = [\varepsilon^{-1}]$. An easy calculation involving (4.21), (4.22) reveals that g^ε, ψ^ε are Fréchet differentiable on H with the Fréchet differentials ∇g^ε and $\nabla\psi^\varepsilon$ Lipschitzian on H, i.e., $g^\varepsilon, \psi^\varepsilon \in C^1_{\mathrm{Lip},\infty}(H)$. Moreover, for every $r > 0$ there exists C_r independent of ε such that

$$|\nabla g^\varepsilon(y)| + |\nabla\psi^\varepsilon(y)| \leq C_r \text{ for } |y| \leq r. \tag{4.23}$$

Arguing as above we may infer that for every $\varepsilon > 0$, Problem (4.18) has at least one optimal pair $(y_\varepsilon, u_\varepsilon)$; we have

$$\int_0^T (g^\varepsilon(y_\varepsilon) + \Phi^*(u_\varepsilon) + \frac{1}{2}|u_\varepsilon - u^*|_U^2)dt + \psi^\varepsilon(y_\varepsilon(T)) \tag{4.24}$$
$$\leq \int_0^T (g^\varepsilon(z_\varepsilon) + \Phi^*(u^*))dt + \psi^\varepsilon(z_\varepsilon(T))$$

where $z_\varepsilon \in W^{1,2}(0,T;H)$ is the solution to

$$z'_\varepsilon - Az_\varepsilon + \beta^\varepsilon(z_\varepsilon) = Bu^*$$

$$z_\varepsilon(0) = y_0.$$

Inasmuch as $z_\varepsilon \to y^*$ in $C([0,T];H)$ for $\varepsilon \to 0$ (see for instance [3]) we have

$$\lim_{\varepsilon \downarrow 0} \int_0^T g^\varepsilon(z_\varepsilon)dt = \int_0^T g(y^*)dt \tag{4.25}$$

and

$$\lim_{\varepsilon \downarrow 0} \psi^\varepsilon(z_\varepsilon(T)) = \psi(y^*(T)). \tag{4.26}$$

By (4.24) we see that on some subsequence, $u_\varepsilon \to u_1$ weakly in $L^2(0,T;H)$ and $y_\varepsilon \to y(t,y_0,u_1)$ weakly in $W^{1,2}(0,T;H)$ and strongly in $C([0,T];H)$. Since the functional $u \to \int_0^T \Phi^*(u)dt$ is weakly lower semicontinuous on $L^2(0,T;U)$ we have

$$\liminf_{\varepsilon \downarrow 0} \int_0^T \Phi^*(u_\varepsilon)dt \geq \int_0^T \Phi^*(u_1)dt.$$

Then by (4.24), (4.25) and (4.26) it follows that $u_1 = u^*$ and

$$u_\varepsilon \to u^* \text{ strongly in } L^2(0,T;U) \tag{4.27}$$

$$y_\varepsilon \to y^* \text{ weakly in } W^{1,2}(0,T;H) \cap L^2(0,T;D(A)) \tag{4.28}$$

and strongly in $C([0,T];H)$.

Let $p_\varepsilon \in W^{1,2}(0,T;H) \cap L^2(0,T;D(A))$ be the solution to boundary value problem

$$p_\varepsilon' + Ap_\varepsilon - p_\varepsilon \nabla\beta^\varepsilon(y_\varepsilon) = \nabla g^\varepsilon(y_\varepsilon) \text{ a.e. } t \in]0,T[\tag{4.29}$$

$$p_\varepsilon(T) = -\nabla\psi^\varepsilon(y_\varepsilon(T)). \tag{4.30}$$

Since u_ε is optimal in Problem (4.18), after some calculation we find that ($\Phi^{*\prime}$ is the directional derivative of Φ^*)

$$\int_0^T (B^*p_\varepsilon + u^* - u_\varepsilon, v)_U dt \leq \int_0^T \Phi^{*\prime}(u_\varepsilon, v)dt \quad \forall v \in L^2(0,T;U).$$

This yields (see §2 in Chapter 1)

$$B^*p_\varepsilon \in \partial\Phi^*(u_\varepsilon) + u_\varepsilon - u^* \quad \text{a.e. } t \in]0,T[. \tag{4.31}$$

By Eqs (4.29), (4.30) we find the estimates

$$|p_\varepsilon(t)|^2 + \int_0^T |p_\varepsilon(t)|_V^2 dt + |\nabla\beta^\varepsilon(y_\varepsilon)p_\varepsilon|_{L^1(Q)} \leq C \quad \forall\varepsilon > 0 \tag{4.32}$$

where $V = D((-A)^{1/2})$ and V^* is the dual of V. In particular we may infer that $\{p'_\varepsilon\}$ is bounded in $L^1(0,T;L^1(\Omega) + V^*)$ and on some subsequence

$$\begin{aligned} &p_\varepsilon \to p \text{ weakly in } L^2(0,T;V) \\ &Ap_\varepsilon \to Ap \text{ weakly in } L^2(0,T;V^*). \end{aligned} \tag{4.33}$$

On the other hand, it follows by assumption (4.5) that

$$\int_E |p_\varepsilon|\ |\nabla\beta^\varepsilon(y_\varepsilon)|dx\ dt \leq C \int_E |p_\varepsilon|\ |\beta^\varepsilon(y_\varepsilon)\ dx\ dt$$

$$+ C\int_E |y_\varepsilon|\ |p_\varepsilon|\ dx\ dt + C\int_E |p_\varepsilon|\ dx\ dt$$

where E is an arbitrary measurable subset of Q. Along with the Dunford-Pettis criterion the latter implies that $\{p_\varepsilon\nabla\beta^\varepsilon(y_\varepsilon)\}$ is a weakly compact subset of $L^1(Q) \subset L^1(0,T;(H^s(\Omega))^*)$ where $s > N/2$. Hence there exists $\mu_p \in L^1(Q)$ such that

$$p_\varepsilon\nabla\beta^\varepsilon(y_\varepsilon) \to \mu_p \text{ weakly in } L^1(Q) \tag{4.34}$$

and therefore

$$p'_\varepsilon \to p' \quad \text{weakly in } L^1(0,T;Y^*) \tag{4.35}$$

where $Y^* = V^* + H^s(\Omega))^*$. Hence $p \in AC([0,T];Y^*)$ and by the Ascoli-Arzelà theorem

$$p_\varepsilon(t) \to p(t) \text{ uniformly on } [0,T] \tag{4.36}$$

in the strong toplogy Y^*.

Recalling that H is compactly imbedded in Y^* (by condition (4.6)) it follows by (4.32), (4.33) and (4.36) that

$$p_\varepsilon \to p \text{ strongly in } L^2(Q). \tag{4.37}$$

Letting ε tend to zero in Eqs (4.29), (4.30) and (4.31) we see that

$$p' + Ap - \mu_p = \xi \quad \text{a.e.} \quad t \in]0,T[$$

$$p(T) = -\eta \qquad (4.38)$$

$$B^*p \in \partial\Phi^*(u^*) \qquad \text{a.e.} \quad t \in]0,T[$$

where η is the weak limit of $\{\nabla\psi^\varepsilon(y_\varepsilon(T))\}$ in H and ξ is weak-star limit in $L^\infty(0,T;H)$ of $\{\nabla g^\varepsilon(y_\varepsilon)\}$. To conclude the proof it remains to show that

$$\mu_p(x,t) \in \partial\beta(y^*(x,t))p(x,t) \quad \text{a.e. } (x,t) \in Q \qquad (4.39)$$

$$\eta \in \partial\psi(y^*(T)); \; \xi(t) \in \partial g(y^*(t)) \quad \text{a.e. } t \in]0,T[. \qquad (4.40)$$

To this end we need the following lemma:

LEMMA 7. *Let* $\{y_n\}$ *be a sequence such that* $y_n \to y$ *strongly in* H *and* $\nabla\phi^\varepsilon(y_n) \to \zeta$ *weakly in* H *for* $n = [\varepsilon^{-1}] \to \infty$. *Then*

$$\zeta \in \partial\phi(y) \qquad (4.41)$$

Here ϕ *is a locally Lipschitz function on* H *and* ϕ^ε *is defined by Eq. (4.22).*

Proof By the theorem of mean we have

$$\lambda^{-1}(\phi^\varepsilon(y_n + \lambda z) - \phi^\varepsilon(y_n)) = \lambda^{-1}(\phi(P_n(y_n + \lambda z) - \varepsilon\Lambda_n\tau_{n,\lambda})$$

$$- \phi(P_n y_n - \varepsilon\Lambda_n\tau_{n,\lambda})$$

where $|\tau_{n,\lambda}|_n \leq 1$. On some subsequence $\lambda \to 0$ we have $\tau_{n,\lambda} \to \tau_n$ and therefore

$$(\nabla\phi^\varepsilon(y_n),P_n z) \leq \phi^0(P_n y_n - \varepsilon\Lambda_n\tau_n, P_n z) \quad \forall z \in H.$$

Since the function ϕ^0 is upper semicontinuous on H × H the latter yields

$$(\zeta,z) \leq \phi^0(y,z) \text{ for all } z \in H.$$

By definition of the generalized gradient $\partial\phi$ the latter implies (4.41) as claimed.

By Lemma 7 and (4.28) we find Eqs (4.40), and Eq. (4.39) follows by the following argument. By Egorov's theorem, for each $\eta > 0$ there exists a measurable subset $E_\eta \subset Q$ such that $\{y_\varepsilon\}$ is bounded on E_η and it is uniformly convergent to y^* for $\varepsilon \to 0$. Thus there exists $f \in L^\infty(E_\eta)$ such that

$$\nabla\beta^\varepsilon(y_\varepsilon) \to f \text{ weak star in } L^\infty(E_\eta).$$

Then by Lemma 7 we may infer via a selection type argument that $f(t,x) \in \partial\beta(y^*(t,x))$ a.e. $(t,x) \in E_\eta$. Since $\{p_\varepsilon\}$ is strongly convergent in $L^2(Q)$ the latter yields (4.39) as claimed.

PROPOSITION 2. *In addition to hypotheses of Proposition 1 assume that condition (v) holds. Then* $p' - Ap \in L^2(Q)$, $p \in C([0,T];H)$ *and*

$$p \in L^2(\delta,T;D(A)),\ p' \in L^2(\delta,T;D(A)) \tag{4.42}$$

for every $\delta \in]0,T[$. *If* $\psi = 0$ *then* $p \in L^2(0,T;D(A)) \cap W^{1,2}(0,T;H)$.

Proof If β is globally Lipschitz then $\nabla\beta^\varepsilon$ is uniformly bounded in $\mathbb{R}$ and so $\{p_\varepsilon \nabla\beta^\varepsilon(y_\varepsilon)\}$ is weakly compact in $L^2(Q)$. Then (4.42) follows by Proposition 9 in Chapter 1. If $\psi = 0$ then $p(T) = 0 \in D(A)$ and $p \in L^2(0,T;D(A)) \cap W^{1,2}(0,T;H)$.

Assume now that $D(A) \subset H^2(\Omega)$ and $N = 1$. Then

$$W^{1,2}(0,T;H) \cap L^2(0,T;D(A)) \subset C(\bar{Q})$$

and by (4.28) we see that $\{y_\varepsilon\}$ is a bounded subset of $C(\bar{Q})$. Hence $\{\nabla\beta^\varepsilon(y_\varepsilon)\}$ is bounded in $C(\bar{Q})$ and the conclusion follows.

The function p arising in Propositions 1 and 2 is called a *dual extremal arc*.

Let $\phi:[0,T] \times H \to \mathbb{R}$ be the optimal value function corresponding to Problem (P), i.e.,

$$\phi(t,y_0) = \inf\left\{\int_t^T (g(y(s,t,y_0,u)) + \Phi^*(u(s))ds + \psi(y(T,t,y_0,u))\right\} \tag{4.43}$$

$$\text{for } t \in [0,T],\ y_0 \in H$$

where $y(s,t,y_0,u)$ is the solution to

$$\begin{aligned} &y'(s) + Ly(s) = Bu(s) \quad \text{a.e. } s \in]t,T[\\ &y(t) = y_0. \end{aligned} \tag{4.44}$$

By Proposition 9 for every $y_0 \in \overline{D(L)} = H$, $y(\cdot,t,y_0,u) \in C([0,T];H)$ and $(s-t)^{1/2}y \in W^{1,2}(0,T;H)$. In particular this implies that the map $u \to y(\cdot,t,y_0,u)$ is compact from $L^2(t,T;U)$ to $C(]t,T[;H \cap L^2(t,T;H)$. Hence for every $(t,y_0) \in [0,T] \times H$ the infimum defining $\phi(t,y_0)$ is attained.

Moreover, we have the obvious inequalities

$$|y(s,t,y_0,u) - y(s,t,\tilde{y}_0,u)| \leq |y_0 - \tilde{y}_0|, \quad t \leq s \leq T$$
$$|y(s,t,y_0,u)| \leq |y_0| + |B|\int_t^s |u(z)|_U dz \quad t \leq s \leq T \tag{4.45}$$

which follow by Eq. (4.44) using the fact that the operator L is monotone in H.

Let $y_0 \in H$ be fixed such that $|y_0| \leq r$. By definition of ϕ we have

$$\phi(t,y_0) \leq \int_t^T (g(y(s,t,y_0,0) + \Phi^*(0))ds$$
$$+ \psi(y(T,t,y_0,0) \leq C_r \tag{4.46}$$

where C_r is a positive constant independent of t. Along with (4.13) the latter inequality implies that in (4.43) we may restrict ourselves to the subset $M_r \subset L^2(t,T;U)$ defined by

$$M_r = \{u \in L^2(t,T,U), \int_t^T |u(s)|_U^2 ds \leq C_r^1\}$$

for some positive constant C_r^1. By (4.45) we see that, for every $u \in M_r$, the function $y_0 \to g(y(s,t,y_0,u)$ is Lipschitzian on the closed ball $\{y_0, |y_0| \leq 1\}$ with Lipschitz constant independent of t. Hence for every $t \in [0,T]$ the function $y \to \phi(t,y)$ is locally Lipschitzian with the Lipschitz constant (on bounded subsets) independent of t.

Now we fix $y_0 \in D(L)$ and consider $\tilde{t} \leq t \leq s \leq T$. By Eq. (4.44) we have

$$|y(s,\tilde{t},y_0,u) - y(s,t,y_0,u)| \leq |y(t,\tilde{t},y_0,u) - y_0| \tag{4.47}$$
$$\leq \int_{\tilde{t}}^t (|Bu(z)| + |Ly_0|)dz.$$

Let $u^t \in L^2(t,T;U)$ and $y^t = y(s,t,y_0,u^t)$ be such that

$$\phi(t,y_0) = \int_t^T (g(y^t(s)) + \Phi^*(u^t(s)))ds + \psi(y^t(T))$$

and let $v(s) = u_0$ for $\tilde{t} \leq s \leq t$, $v(s) = u^t(s)$ for $t \leq s \leq T$. We have

$$\phi(\tilde{t},y_0) - \phi(t,y_0) \leq \int_{\tilde{t}}^{t} (g(s,\tilde{t},y_0,v) + \Phi^*(u_0))ds$$

$$+ \int_t^T (g(y(s,\tilde{t},y_0,v)) - g(s,t,y_0,v))ds + \psi(y(T(\tilde{t},y_0,v))$$

$$- \psi(y(T,t,y_0,v))).$$

Then by (4.47) we see that

$$|\phi(\tilde{t},y_0) - \phi(t,y_0)| \leq C|\tilde{t} - t| \text{ for all } t, \tilde{t} \in [0,T].$$

We have therefore proved the following lemma:

LEMMA 8. *For every* $t \in [0,T], \phi(t,\cdot)$ *is locally Lipschitz and for every* $y_0 \in D(L)$ *the function* $t \to \phi(t,y_0)$ *is Lipschitz on* $[0,T]$.

We notice also

LEMMA 9. *For all* $y_0 \in H$ *and* $t \in [0,T]$ *we have*

$$\phi(0,y_0) = \inf \{\int_0^t (g(y(s,0,y_0,u)) + \Phi^*(u(s)))ds$$
$$+ \phi(t,y(t,0,y_0,u)); u \in L^2(0,t;U)\} \qquad (4.48)$$

Proof Let (y,u) be such that $y = y(s,0,y_0,u)$ and

$$\phi(0,y_0) = \int_0^t (g(y(s)) + \Phi^*(u(s)))ds + \int_t^T (g(y(s))$$

$$+ \Phi^*(u(s)))ds + \psi(y(T)).$$

This yields

$$\phi(0,y_0) \geq \phi(t,y(t)) + \int_0^t (g(y(s)) + \Phi^*(u(s)))ds. \qquad (4.49)$$

On the other hand, for all $u \in L^2(0,T;U)$ and $y = y(s,0,y_0,u)$ we have

$$\phi(0,y_0) \leq \int_0^t (g(y(s)) + \Phi^*(u(s)))ds + \int_t^T (g(y(s))$$

$$+ \Phi^*(u(s)))ds + \psi(y(T)).$$

Clearly, we may choose the pair (y,u) in such a way that

$$\phi(t,y(t)) = \int_t^T (g(y(s)) + \Phi^*(u(s)))ds + \psi(y(T))$$

and therefore

$$\phi(0,y_0) \leq \int_0^t (g(y(s)) + \Phi^*(u(s)))ds + \psi(t,y(t)).$$

Along with (4.49) the latter yields (4.48) as claimed.

Proposition 3 below amounts to saying that under assumptions (i)-(iv) the feedback law

$$u = \nabla\Phi(D\partial\phi(t,y)), \quad t \in [0,T], \quad y \in H \tag{4.50}$$

is an *optimal feedback control* of Problem (P).

PROPOSITION 3. *Assume that conditions (i)-(iv) are satisfied. Let* (y^*,u^*) *be an optimal pair Problem* (P) *where* $y_0 \in D(1)$. *Then*

$$u^*(t) \in \nabla\phi(-B^*\partial\phi(t,y^*(t))) \quad \text{a.e.} \quad t \in]0,T[. \tag{4.51}$$

If in addition conditions (v) and (vi) hold then u^* *is continuous and (4.51) extends to all* $t \in [0,T]$.

Proof ([4]). By Lemma 9, for every $t \in [0,T]$ the pair (y^*,u^*) is optimal on $[0,t]$ for the control problem

$$\inf \{\int_0^t (g(y(s,0,y_0,u) + \Phi^*(u(s)))ds + \phi(t,y(t,0,y_0))\}. \tag{4.52}$$

Hence by Proposition 1, for every $t \in [0,T]$ there exists $p^t \in AC([0,t];Y^*) \cap C_w([0,t];H)$ such that

$$B^*p^t(s) \in \partial\phi^*(u^*(s)) \quad \text{a.e.} \quad s \in]0,t[\tag{4.53}$$

$$p^t(t) \in -\partial\phi(t,y^*(t)). \tag{4.54}$$

Let F be the set of all points t where u^* is approximatively continuous, i.e., there exists a measurable subset $F_t \subset [0,T]$ such that t is a density point for F_t and u is continuous on F_t at t. It is well known that the

Lebesgue measure of F is T.

Let F^t be the set of all $s \in [0,t]$ for which (4.53) is satisfied, t being a point of F. Obviously there exists at least one sequence $\{t_n\} \subset F_t \cap F^t$ convergent to t for $n \to \infty$. Hence $u^*(t) = \lim_{n\to\infty} u^*(t_n)$ where $B^*p^t(t_n) \in \partial\phi^*(u^*(t_n))$. Since $p^t(t_n)$ is weakly convergent to $p^t(t)$ and $\partial\Phi^*$ is strongly weakly closed in U × U we may infer that Eq. (4.53) holds for all $s,t \in F$. Along with (4.54) this implies (4.51) as claimed.

If condition (v) holds then by Proposition 2, we may choose $p^t \in C([0,t],H)$ and by (4.53) we see that

$$u^*(s) = \nabla\Phi(B^*p^t(s)) \quad \text{a.e. } s \in]0,t[.$$

Hence u^* is a continuous function on [0,t] and Eq. (4.51) extends to all of [0,t] thereby completing the proof.

For all (t,y) in $[0,T] \times D(L)$ define

$$D_t\phi(t,y) = \{w \in \mathbb{R}: \exists\, y_\lambda \xrightarrow{L} y,\ \lambda \to 0 \text{ such that} \\ \lambda^{-1}(\phi(t+\lambda,y_\lambda) - \phi(t,y_\lambda)) \to w\} \tag{4.55}$$

where we have used the notation: $y_\lambda \xrightarrow{L} y \rightleftarrows y_\lambda \to y$ and $Ly_\lambda \to Ly$ strongly in H. By Lemma 8 (see (4.47), (4.48)) it follows that for all $(t,y) \in [0,T] \times D(L)$, $D_t\phi(t,y)$ is a nonempty bounded subset of $\mathbb{R}$.

Now we are ready to formulate the main existence theorem for the backward value problem associated with Hamilton-Jacobi equation (4.1).

THEOREM 4. *Under assumptions (i)-(vi) the function* ϕ *defined by (4.43) where* $\psi = 0$ *satisfies the equation*

$$D_t\phi(t,y) - \Phi(D\partial\phi(t,y)) - (Ly,\partial\phi(t,y)) + g(y) = 0 \tag{4.56}$$

$$\textit{for all } (t,y) \in [0,T] \times D(L)$$

and the final value condition

$$\phi(T,y) = 0 \textit{ for all } y \textit{ in } H \tag{4.57}$$

in the following sense: for all (t,y) *in* $[0,T] \times D(L)$ *there exist* $\zeta(t,y)$, $\eta(t,y) \in H$ *and* $\delta(t,y) \in \mathbb{R}$ *such that*

$$\eta(t,y),\ \zeta(t,y) \in \partial\phi(t,y);\ \delta(t,y) \in D_t\phi(t,y) \tag{4.58}$$

and

$$\begin{aligned}&\delta(t,y) - \Phi(D\eta(t,y)) - (Ly,\zeta(t,y)) + g(y)\\ &= (D(\zeta(t,y) - \eta(t,y)),\ \nabla\Phi(D\eta(t,y)))\ \textit{for all}\\ &\quad (t,y) \in [0,T] \times D(L)\end{aligned} \tag{4.59}$$

Proof Let $y_0 \in D(L)$ be arbitrary but fixed. As seen earlier there exist $y^t \in W^{1,2}(t,T;H)$, $u^t \in L^2(t,T;U)$ such that $y^t(s) = y(s,t,y_0,u^t)$ for $s \in [t,T]$ and

$$\phi(t,y_0) = \int_t^T (g(y^t(\tau)) + \Phi^*(u^t(\tau)))d\tau\,.$$

By Lemma 3 it follows that

$$\phi(s,y^t(s)) = \int_s^T (g(y^t(\tau)) + \Phi^*(u^t(\tau)))d\tau \text{ for } t \leqslant s \leqslant T. \tag{4.60}$$

Now by Proposition 2 the dual extremal arc p^t associated with the optimal pair (y^t,u^t) belongs to $W^{1,2}(t,T;H)$. Since

$$u^t(s) = \nabla\Phi(B^*p^t(s)) \text{ for } s \in [t,T] \tag{4.61}$$

and $\nabla\Phi$ is locally Lipschitz on H we may infer that

$$u^t \in W^{1,2}(t,T;U). \tag{4.62}$$

On the other hand, by the conjugacy formula

$$\Phi^*(v) + \Phi(v^*) = (v,v^*)_U \text{ for } v = \nabla\Phi(v^*) \tag{4.63}$$

we have

$$\Phi^*(u^t(s)) + \Phi(B^*p^t(s)) = (u^t(s),B^*p^t(s))_U \tag{4.64}$$

$$\text{for } s \in [t,T].$$

By (4.62) and (4.64) we see that the function $s \to \Phi^*(u^t(s)))$ is continuous on $[t,T]$. Thus by (4.60) we may conclude that the function $s \to \phi(s,y^t(s))$ is continuously differentiable on [t,T] and

$$\frac{d}{ds}\,\phi(s,y^t(s))) = -\,g(y^t(s)) - \Phi^*(u^t(s)) \tag{4.65}$$

for all $s \in [t,T]$.

Next by Proposition 3

$$u^t(s) \in \nabla\Phi(-B^*\partial\phi(s,y^t(s))) \text{ for all } s \in [t,T]. \tag{4.66}$$

On the other hand, since $Bu^t \in W^{1,2}(t,T;H)$ we may infer (see Proposition 9 in Chapter 1) that $\frac{d^+}{ds}\,y^t(s)$ exists at all $s \in [t,T]$ and the function $s \to Ly^t(s)$ is continuous from the right on [t,T]. We have

$$\frac{d}{ds}\,\phi(s,y^t(s))(t) = \lim_{\varepsilon \downarrow 0} \varepsilon^{-1}(\phi(t+\varepsilon,y^t(t+\varepsilon)) - \phi(t,y^t(t+\varepsilon)))$$

$$+ \lim_{\varepsilon \downarrow 0} (\phi(t,y^t(t+\varepsilon)) - \phi(t,y^t(t)))\varepsilon^{-1} = \delta(t,y_0)$$

$$+ \lim_{\varepsilon \downarrow 0} (\zeta_\varepsilon, y^t(t+\varepsilon) - y^t(t))\varepsilon^{-1}$$

where $\delta(t,y_0) \in D_t\phi(t,y_0)$ and $\zeta_\varepsilon \in \partial\phi(t,\theta_\varepsilon)$, θ_ε is a point on the segment between $y^t(t+\varepsilon)$ and $y^t(t)$ (see Proposition 7 part (iv) in Chapter 1). Selecting a subsequence if necessary, we may assume that ζ_ε is weakly convergent to $\zeta(t,y_0)$ for $\varepsilon \to 0$. Recalling that the map $\partial\phi: H \to H$ is w*-upper semicontinuous we conclude that $\zeta(t,y_0) \in \partial\phi(t,y_0)$ and

$$\frac{d^+}{ds}\,\phi(t,y^t(t)) = \delta(t,y_0) + (\zeta(t,y_0), \frac{d^+}{ds}\,y^t(t)).$$

Let $\eta(t,y_0) \in \partial\phi(t,y_0)$ be such that (see (4.66))

$$u^t(t) = \nabla\Phi^*(-B^*\eta(t,y_0)).$$

Recalling that

$$\frac{d^+}{ds}\,y^t(t) = Bu^t(t) - Ly^t(t) = B\nabla\Phi^*(-B^*\eta(t,y_0)) - Ly_0$$

we see that

$$\delta(t,y_0) + (\zeta(t,y_0), B\nabla\Phi^*(-B^*\eta(t,y_0)) - Ly_0)$$

$$+ g(y_0) + \Phi^*(\nabla\Phi^*(-B^*\eta(t,y_0))) = 0.$$

Using once again the conjugacy formula (4.63) and substituting $B = -D^*$ in the latter equation we get Eq. (4.59). Since y_0 is arbitrary in $D(L)$ the proof of Theorem 4 is complete.

REMARK 3. 1º If $\partial\phi(t,y)$ happens to be single-valued at (t,y) then we see by (4.58) and (4.59) that ϕ satisfies Eq. (4.56).

One may suspect that the set of those points (t,y) is dense in $[0,T] \times D(L)$.

2º Assume that g,ψ and β are differentiable with the differentials locally bounded. Let (y^t,u^t) be an optimal pair of Problem (P) on the interval $[t,T]$, i.e.,

$$\phi(t,y_0) = \int_t^T (g(y^t(s)) + \Phi^*(u^t(s)))ds + \psi(y^t(T)) \tag{4.67}$$

and let $p^t:[0,T] \to H$ be the corresponding dual extremal arc, i.e.,

$$\begin{aligned}&(y^t)' + Ly^t = B\nabla\Phi(B^*p^t) \quad \text{a.e. } s \in]t,T[.\\ &(p^t)' + Ap^t - p^t\partial\beta(y^t) = \nabla g(y^t) \text{ a.e. } s \in [t,T].\end{aligned} \tag{4.68}$$

$$y^t(t) = y_0, p^t(T) = -\nabla\psi(y^t(T)) \tag{4.69}$$

and by Corollary 1

$$p^t(s) \in - \partial\phi(s,y^t(s))) \text{ for all } s \in [t,T]. \tag{4.70}$$

Let (z^t,v^t,q^t) be the solution to Eqs (4.68), (4.69) corresponding to z_0. A little calculation involving system (4.68) and Gronwall's lemma gives

$$|y^t(s) - z^t(s)| + |p^t(s) - q^t(s)| \leq C|y_0 - z_0|$$

$$\text{for } s \in [T-\delta,T]$$

where δ is sufficiently small. Then a glance at (4.68) reveals that for $t \in [T-\delta,T]$,

$$|\phi(t,y_0) - \phi(t,z_0) + (p^t(t),y_0 - z_0)| \leq o(|y_0 - z_0|).$$

In other words, $\phi(t,\cdot)$ is continuously differentiable on H and therefore it is a classical solution to Eq. (4.1) (by Theorem 1 unique) on $[T-\delta,T]$.

By Lemma 9 we see that for all $t \leqslant T_1 \leqslant T$,

$$\phi(t,y_0) = \int_t^{T_1} (g(y^t(s)) + \Phi^*(u^t(s)))ds + \phi(T_1,y^t(T_1)).$$

Hence, if the function $\phi(T_1,\cdot)$ happens to be differentiable at $y^t(T_1)$, we may infer by the above reasoning that ϕ is a classical solution to Eq. (4.1) on a whole interval $[T_1-\delta,T_1]$. This resembles the classical picture of solutions of Hamilton-Jacobi equations whose derivatives have discontinuities across curves which separate regions in which the solution is smooth. In general, these discontinuities develop even for smooth initial values.

§5 Stationary Hamilton-Jacobi equations and control problems with infinite time horizon

We shall study here the existence for the stationary Hamilton-Jacobi equation

$$\Phi(D\phi_y(y)) + (\phi_y(y),Ly) = g(y) \text{ for all } y \in D(L) \tag{5.1}$$

where the functions $\Phi:U \to \mathbb{R}$, $g:H = L^2(\Omega) \to \mathbb{R}$ and $L.D(L) \subset H \to H$ satisfy assumptions (i)-(iv). Assume further that the following conditions hold:

(j) $g(0) = 0$;Φ^* is Gâteaux differentiable on U and $\Phi^*(0) = 0, \nabla\Phi^*(0) = 0$.

(jj) The range $R(D^*)$ of D^* is dense in H.

(jjj) There exists $\omega > 0$ such that

$$(Ly,y) \geqslant \omega|y|^2 \text{ for all } y \text{ in } H.$$

Here Φ^* denotes as usual the conjugate of Φ and D^* is the adjoint of $D \in \mathcal{L}(H,U)$.

We associate with Eq. (5.1) the infinite time horizon control problem

$$\inf \{\int_0^\infty (g(y(s,0,y_0,u)) + \Phi^*(u(s)))ds; \quad u \in L^2_{loc}(\mathbb{R}^+;U)\} = \phi(y_0) \tag{5.2}$$

where $y(s,0,y_0,u)$ is the solution to the initial value Problem (4.10).

We shall see that ϕ is a solution in some generalized sense to Eq. (5.1). To this end we shall prove firstly the following lemma:

LEMMA 10. *The function* ϕ *is locally Lipschitz on* H *and for every* $y_0 \in H$ *the infimum defining* $\phi(y_0)$ is attained.

Proof Let y_0 be fixed but arbitrary in H. By condition (jjj) there exists $u \in L^2(\mathbb{R}^+;U)$ such that $\Phi^*(u) \in L^1(\mathbb{R}^+)$ and $g(y(t,0,y_0,u)) \in L^1(\mathbb{R}^+)$. Indeed it suffices to take $u = -B^*y$ where y is the solution to

$$y' + Ly + BB^*y = 0, \quad \text{a.e. } t > 0$$

$$y(0) = y_0.$$

Hence $\phi(y_0) < +\infty$. Using the fact that for every y_0 the map $u \to y(t,0,y_0,u)$ is compact from $L^2(0,T;U)$ to $C(]0,T];H)$ it follows by standard arguments that the infimum in (5.2) is attained.

Now for $|y_0| \leq r$ we have by (jjj),

$$\phi(y_0) \leq \int_0^\infty g(y(t,0,y_0,0))dt \leq C_r$$

and therefore

$$\phi(y_0) = \inf \{\int_0^\infty (g(y(s,0,y_0,u)) + \Phi^*(u(s)))ds; u \in U_r\} \tag{5.3}$$

where

$$U_r = \{u \in L^2_{loc}(\mathbb{R}^+;U); \int_0^\infty \Phi^*(u(t))dt \leq C_r\}.$$

In view of condition (ii) (see (4.14)) we see by Eq. (4.10) that

$$|y(t,0,y_0,u)| \leq \exp(-\omega t)|y_0| + |B| \int_0^t \exp(-\omega(t-s))|u(s)|_U ds \tag{5.4}$$

$$\leq \exp(-\omega t)r + |B|C_r^1 \leq C_r^2, \quad \forall u \in U_r.$$

Let $y_0, z_0 \in \Sigma_r = \{y \in H; |y| \leq r\}$. We have

$$\phi(y_0) - \phi(z_0) \leq \int_0^\infty (g(y(t,0,y_0,v^*)) - g(y(t,0,z_0,v^*)))dt$$

where v^* is such that

$$\phi(z_0) = \int_0^\infty (g(y(t,0,z_0,v^*)) + \Phi^*(v(t)))dt.$$

Since g is locally Lipschitz and by (jjj)

$$|y(t,0,y_0,v^*) - y(t,0,z_0,v^*)| \leq \exp(-\omega t)|y_0 - z_0|, \quad t \geq 0$$

we see by (5.4) that

$$\phi(y_0) - \phi(z_0) \leq L_r|y_0 - z_0| \text{ for all } y_0,z_0 \in \Sigma_r$$

thereby completing the proof.

In particular, it follows by Lemma 10 that for every $y_0 \in H$ the control problem has at least one optimal pair $(y^*,u^*) \in C(\mathbb{R}^+;H) \times L^2_{loc}(\mathbb{R}^+;U)$. We shall assume from now on that $y_0 \in D(1)$, i.e., y_0 satisfies (4.11). Then

$$y^* = y(t,0,y_0,u^*) \in W^{1,2}_{loc}(\mathbb{R}^+;H) \cap L^2_{loc}(\mathbb{R}^+;D(A)).$$

Proposition 4 below is a maximum principle type result for Problem (5.1) and has an intrinsic interest.

PROPOSITION 4. *Assume that conditions (i)-(iv) and (j)-(jjj) are satisfied. If* (y^*,u^*) *is an optimal pair of Problem (5.1) then there exists a function* $p \in L^\infty(\mathbb{R}^+;H) \cap L^2_{loc}(\mathbb{R}^+;V) \cap AC_{loc}(\mathbb{R}^+;Y^*) \cap C_w(\mathbb{R}^+;H)$ *satisfying the system*

$$p' + Ap - p\,\partial\beta(y^*) - \partial g(y^*) \ni 0 \quad \text{a.e.} \quad t > 0 \tag{5.5}$$

$$B^*p(t) = \nabla\Phi^*(u^*(t)), \quad \text{a.e.} \quad t > 0 \tag{5.6}$$

$$p(t) + \partial\phi(y^*(t)) \ni 0 \quad \textit{for all} \quad t \geq 0. \tag{5.7}$$

Proof Arguing as in the proof of Lemma 9 we see that

$$\phi(y_0) = \inf\{\int_0^t (g(y(s,0,y_0,u)) + \Phi^*(u(s)))ds + \phi(y(t)),\ u \in L^2(0,t;U)\}. \tag{5.8}$$

Thus by Proposition 1 there exists a function

$$p^t \in L^\infty(0,t;H) \cap L^2(0,t;V) \cap AC([0,t];Y) \cap C_w([0,t];H)$$

such that

$$(p^t)'(s) + Ap^t(s) - p^t(s)\partial\beta(y^*(s)) \tag{5.9}$$

$$- \partial g(y^*(s)) \ni 0 \quad \text{a.e. } s \in]0,t[$$

$$p^t(t) + \partial\phi(y^*(t)) \ni 0 \tag{5.10}$$

$$B^*p^t(s) = \nabla\Phi^*(u^*(s)) \quad \text{a.e. } s \in]0,t[. \tag{5.11}$$

Since by assumptions (jj) the kernel $N(B^*)$ of B^* consists only of $\{0\}$, it follows by (5.11) that

$$p^t(s) = p^{t'}(s) \text{ for } 0 \leqslant s \leqslant t \leqslant t'.$$

Let $p : [0,\infty[\to H$ be the function defined by

$$p(s) = p^t(s) \text{ for } s \in [0,t].$$

Clearly p satisfies Eqs (5.5)-(5.7). On the other hand, since $\Phi^*(u^*) \in L^1(\mathbb{R}^+)$ we see by (4.14) that $u^* \in L^2(\mathbb{R}^+;U) + L^\infty(\mathbb{R}^+;U)$ and again by condition (jjj) we may infer that $y^* \in L^\infty(\mathbb{R}^+;H)$. Since $\partial\phi$ is locally bounded the latter implies that $p \in L^\infty(\mathbb{R}^+;H)$ thereby completing the proof.

Coming back to Eq. (5.8) we see that it yields

$$\phi(y^*(t)) = \int_t^\infty (g(y^*(s)) + \Phi^*(u^*(s)))ds \text{ for all } t \geqslant 0$$

and therefore

$$\frac{d}{dt}\phi(y^*(t)) + g(y^*(t)) + \Phi^*(u^*(t)) = 0 \quad \text{a.e. } t > 0. \tag{5.12}$$

Using once again the mean value theorem we see that

$$\frac{d}{dt}\phi(y^*(t)) = (\zeta_0(t), (y^*)'(t)) \quad \text{a.e. } t > 0 \tag{5.13}$$

where $\zeta_0(t) \in \partial\phi(y^*(t))$.

Then combining Eqs (5.6), (5.7), (5.12) and (5.13) we get

$$(\zeta_0(t), B\nabla\Phi^*(-B^*\eta_0(t)) - Ly^*(t)) + g(y^*(t))$$
$$+ \Phi^*(\nabla\Phi(-B^*\eta_0(t))) = 0 \quad \text{a.e. } t > 0$$

where $\eta_0(t) \in \partial\phi(y^*(t))$. By the conjugacy formula (4.63) we see that

$$\Phi^*(-B^*\eta_0(t)) + (Ly^*(t),\zeta_0(t)) = g(y^*(t)) - (B^*(\eta_0(t) - \zeta_0(t)),$$
$$\nabla\Phi(-B^*\eta_0(t))).$$

Since y^* is continuous and $y^*(0) = y_0$ is any element of D(L) we conclude that there exists a dense subset E of D(L) such that

$$\Phi(D\eta(y)) + (\zeta(y),Ly) = g(y) - (D(\zeta(y)) - \eta(y)), \tag{5.14}$$
$$\nabla\Phi(D\eta(y)) \text{ for all } y \in E$$

where

$$\eta(y),\zeta(y) \in \partial\phi(y) \text{ for all } y \in E\ . \tag{5.15}$$

We have therefore proved

THEOREM 5. *Under assumptions (i)-(iv) and (j)-(jjj) the function ϕ given by (5.2) satisfies Eq. (5.1) in the sense of Eqs (5.14), (5.15).*

It is useful to notice that by virtue of Eqs (5.5)-(5.7) any optimal control u^* of Problem (5.2) is a feedback control of the form

$$u^*(t) = \nabla\Phi(D\partial\phi(y^*(t))) \quad \text{a.e.} \quad t > 0. \tag{5.16}$$

Time optimal control problems

We shall indicate now some applications of the previous results to time optimal control problems of the form:

$$\inf \{T; y'(t) + Ly(t) = u(t); |u(t)| \leq 1, \tag{5.17}$$
$$\text{a.e. } t \in]0,T[\quad y(0) = y_0,\ y(T) = 0\} = T(y_0)$$

under assumptions (i)-(iv) and (j)-(jjj).

It is easily seen that for every $y_0 \in H$ there exists at least one optimal control u^* in Problem (5.17). Such a control u^* is called *time optimal control* and the corresponding time $T(y_0)$ is called *optimal time*.

For the general theory of time optimal control problems associated with infinite dimensional linear systems we refer the reader to Lions' book [14] and to Fattorini's work [12]. Here we shall indicate an approximating method to study Problem (5.17).

Let $\alpha \in C^\infty(\mathbb{R}^+)$ be such that $0 \leq \alpha \leq 1$ and

$$\alpha(r) = \begin{cases} 1 & \text{for} \quad r \geq 2 \\ 0 & \text{for} \quad 0 \leq r \leq 1. \end{cases}$$

Define the function $g^\varepsilon : H \to \mathbb{R}$ by

$$g^\varepsilon(y) = \alpha(|y|^2/\varepsilon^2) \text{ for all } \varepsilon > 0 \text{ and } y \in H \tag{5.18}$$

and set

$$G^\varepsilon = \nabla g^\varepsilon, \; h^\varepsilon(x) = (2\varepsilon)^{-1}((|u|-1)^+)^2 \text{ for all } u \in H. \tag{5.19}$$

The reader will note that h^ε is the regularization

$$h^\varepsilon(u) = \inf \{(2\varepsilon)^{-1}|u - v|^2; \; |v| \leq 1\}$$

of the indicator function associated with the constrained control set $\{u; \; |u| \leq 1\}$.

Then consider the infinite time horizon control problem

$$\begin{aligned} \phi^\varepsilon(y_0) = \inf \{&\int_0^\infty (g^\varepsilon(y(s,0,y_0,u)) + h^\varepsilon(u(s)))ds; \\ &u \in L^2_{loc}(\mathbb{R}^+;H)\}. \end{aligned} \tag{5.20}$$

We notice that assumptions (i)-(iv), (j)-(jjj) are trivially satisfied in this case where $g = g^\varepsilon$, $B = 1$ (the identity operator) and

$$\Phi(p) = (h^\varepsilon)^*(p) = |p| + \varepsilon/2|p|^2 \text{ for all } p \in H. \tag{5.21}$$

Hence, for every $\varepsilon > 0$, Problem (5.20) has at least one optimal pair $(y^\varepsilon, u^\varepsilon) \in W^{1,2}_{loc}(\mathbb{R}^+;H) \times L^2_{loc}(\mathbb{R}^+;H)$ and by Proposition 4 there exists $p^\varepsilon \in L^\infty(\mathbb{R}^+,H) \cap L^2_{loc}(\mathbb{R}^+,V) \cap AC_{loc}(\mathbb{R}^+;Y^*) \cap C_w(\mathbb{R}^+;H)$ which satisfies along with y^ε and u^ε the optimality system

$$\begin{aligned} &(y^\varepsilon)' + Ly^\varepsilon = u^\varepsilon \quad \text{a.e. } \; t > 0 \\ &(p^\varepsilon)' + Ap^\varepsilon - p^\varepsilon \partial\beta(y^\varepsilon) - G^\varepsilon(y^\varepsilon) \ni 0 \quad \text{a.e. } t > 0 \end{aligned} \tag{5.22}$$

$$p^{\varepsilon}(t) = \nabla h^{\varepsilon}(u^{\varepsilon}(t)) = \varepsilon^{-1}|u^{\varepsilon}(t)|^{-1}u^{\varepsilon}(t)$$

$$(|u^{\varepsilon}(t)|-1) \quad \text{if} \quad |u^{\varepsilon}(t)| \geq 1 \tag{5.23}$$

$$p_{\varepsilon}(t) = 0 \text{ if } |u^{\varepsilon}(t)| < 1$$

$$p^{\varepsilon}(t) + \partial\phi^{\varepsilon}(y^{\varepsilon}(t)) \ni 0 \text{ for all } t \geq 0. \tag{5.24}$$

By (5.14) and (5.21) we see that ϕ^{ε} is a solution to the Hamilton-Jacobi equation

$$\varepsilon/2|\partial\phi^{\varepsilon}(y)|^2 + |\partial\phi^{\varepsilon}(y)| + (Ly,\partial\phi^{\varepsilon}(y)) = g^{\varepsilon}(y) \tag{5.25}$$

in the sense that there exist $\zeta_{\varepsilon} \subset \partial\phi^{\varepsilon}$ and $\eta_{\varepsilon} \subset \partial\phi^{\varepsilon}$ such that

$$(\zeta_{\varepsilon}(y),Ly + \operatorname{sgn}\eta_{\varepsilon}(y) + \varepsilon\eta_{\varepsilon}(y)) = g^{\varepsilon}(y) + \varepsilon/2|\eta_{\varepsilon}(y)|^2 \tag{5.26}$$

for all y in a dense subset of D(L).

Now let us assume that condition (v) holds. Then as seen in Proposition 2, p^{ε} is locally absolutely continuous on $[0, +\infty[$. Multiplying the first equation in (5.22) by $(p^{\varepsilon})'$ the second by $(y^{\varepsilon})'$ and subtracting the results we get

$$\frac{d}{dt}((p^{\varepsilon}(t),Ly^{\varepsilon}(t)) - |p^{\varepsilon}(t)| - \varepsilon/2|p^{\varepsilon}(t)|^2$$

$$+ g^{\varepsilon}(y^{\varepsilon}(t))) = 0 \quad \text{a.e. } t > 0. \tag{5.27}$$

By (5.23) we see that $p^{\varepsilon} \in L^2(\mathbb{R}^+;H)$ (because $h^{\varepsilon}(u^{\varepsilon}) \in L^1(\mathbb{R}^+)$). On the other hand, recalling that (see (5.16) and (5.21))

$$u_{\varepsilon}(t) = \operatorname{sgn} p^{\varepsilon}(t) + \varepsilon p^{\varepsilon}(t) \quad \text{a.e.} \quad t > 0 \tag{5.28}$$

$$(y^{\varepsilon})' + Ly^{\varepsilon} = u^{\varepsilon} \quad \text{a.e.} \quad t > 0 \tag{5.29}$$

it follows that

$$\int_0^t |Ly^{\varepsilon}(s)|^2 ds \leq C(t + 1) \text{ for all } t \geq 0.$$

Since $g^{\varepsilon}(y^{\varepsilon}) \in L^1(\mathbb{R}^+)$ and $p^{\varepsilon} \in L^2(\mathbb{R}^+;H)$ the latter implies that there exists at least one sequence $t_n \to +\infty$ such that

$$(p^\varepsilon(t_n), Ly^\varepsilon(t_n)) - |p^\varepsilon(t_n)| - \varepsilon/2|p^\varepsilon(t_n)|^2$$

$$+ g^\varepsilon(y^\varepsilon(t_n)) \to 0, \text{ as } n \to \infty .$$

By (5.27) we may conclude therefore that

$$(p^\varepsilon(t), Ly^\varepsilon(t)) - |p^\varepsilon(t)| - \varepsilon/2|p^\varepsilon(t)| + g^\varepsilon(y^\varepsilon(t)) = 0$$

$$\text{for all } t \geq 0$$

and by (5.28) it follows that there exists a single valued section $\eta^\varepsilon \subset \partial\phi^\varepsilon$ such that

$$\varepsilon/2|\eta^\varepsilon(y)|^2 + |\eta^\varepsilon(y)| + (Ly, \eta^\varepsilon(y)) = g^\varepsilon(y) \tag{5.30}$$

for all $y \in E$ where E is a dense subset of $D(L)$.

The relationship between Problems (5.17) and (5.20) becomes clear in Proposition 5 below:

PROPOSITION 5. *Let* y_0 *be fixed in* H. *Then*

$$\lim_{\varepsilon \downarrow 0} \phi^\varepsilon(y_0) = T(y_0) \tag{5.31}$$

and on some subsequence (again denoted ε*) one has*

$$u^\varepsilon \to u^* \textit{ weakly in } L^2_{loc}(\mathbb{R}^+; H) \tag{5.32}$$

$$y^\varepsilon \to y^* \textit{ strongly in } C([0,T], H) \tag{5.33}$$

where u^* *is a time optimal control and* $y^*(t) = y(t,0,y_0,u^*)$ *is the corresponding state.*

Proof For simplicity we shall assume that $y_0 \in D(1) = D((-A)^{1/2})$. Let $T^* = T(y_0)$ and let (y_1^*, u_1^*) be an optimal pair for Problem (5.17). We have

$$\phi^\varepsilon(y_0) \leq \int_0^\infty (g^\varepsilon(y_1^*(t)) + h^\varepsilon(u_1^*(t)))dt$$

$$= \int_0^{T^*} (g^\varepsilon(y_1^*(t)) + h^\varepsilon(u_1^*(t)))dt = \int_0^{T^*} g^\varepsilon(y_1^*(t))dt.$$

Hence

$$\phi^{\varepsilon}(y_0) \leq T^* \text{ for all } \varepsilon > 0. \tag{5.34}$$

The latter implies in particular, that $\{u^{\varepsilon}\}$ is bounded in $L^2_{loc}(\mathbb{R}^+;H)$ and therefore $\{y^{\varepsilon}\}$ is compact in every $C([0,T];H)$. Hence there exists a pair $(y^*,u^*) \in W^{1,2}_{loc}(\mathbb{R}^+;H) \times L^2_{loc}(\mathbb{R}^+;H)$ such that $y^*(t) = y(t,0,y_0,u^*)$ and

$$u^{\varepsilon} \to u^* \text{ weakly in } L^2_{loc}(\mathbb{R}^+;H)$$

$$y^{\varepsilon}(t) \to y^*(t) \text{ uniformly on every } [0,T].$$

Along with (5.20) and (5.34) the latter yields

$$\limsup_{\varepsilon \downarrow 0} \int_0^T g^{\varepsilon}(y^{\varepsilon}(t))dt \leq T^* \text{ for all } T > 0. \tag{5.35}$$

Hence there exist a sequence $\varepsilon_n \to 0$ and T_0 independent of n such that

$$|y^{\varepsilon_n}(t)| \leq 2\varepsilon_n \text{ for } t \geq T_0. \tag{5.36}$$

For, otherwise for every sequence $\{\varepsilon_n\}$ convergent to zero there would exist $t_n \to \infty$ such that

$$|y^{\varepsilon_n}(t_n)| > 2\varepsilon_n \text{ for all } n.$$

Let $\varepsilon_n = n^{-1/2}$. By an easy calculation involving Eq. (5.29) it follows that

$$|y^{\varepsilon_n}(t_n)| \leq |y^{\varepsilon_n}(t)| + \int_t^{t_n} |u^{\varepsilon_n}(s)|ds \text{ for } t \leq t_n. \tag{5.37}$$

Since

$$(2\varepsilon_n)^{-1} \int_0^{\infty} ((|u^{\varepsilon_n}(t)| - 1)^+)^2 dt \leq C$$

it follows by (5.37) that

$$|y^{\varepsilon_n}(t_n)| \leq |y^{\varepsilon_n}(t)| + C(2\varepsilon_n|t - t_n|)^{1/2} \text{ for } t \leq t_n$$

and therefore for a sufficiently large n

$$|y^{\varepsilon_n}(t)| \geq 2^{1/2}\varepsilon_n \text{ for } t \in [t_n - \delta_n, t_n]$$

where $\delta_n = Cn^{-1}$. This would imply that

$$\lim_{n\to\infty} m\{t; |y^{\varepsilon_n}(t)| \geq \sqrt{2}\varepsilon_n\} = +\infty$$

(m is the Lebesgue measure) contrary to (5.36).

Inequality (5.36) implies that $y^*(t) = 0$ for $t \geq T_0$. Let $\tilde{T} = \inf\{T; y^*(T) = 0\}$. We will show that $\tilde{T} = T^*$. To this end, consider the set

$$E_\varepsilon = \{t \in [0,\tilde{T}]; |y^\varepsilon(t)| \geq 2^{1/2}\varepsilon\}.$$

By (5.35)

$$\limsup_{\varepsilon\downarrow 0} m(E_\varepsilon) \leq T^*. \tag{5.38}$$

On the other hand, $\lim_{\varepsilon\downarrow 0} m(E_\varepsilon) = \tilde{T}$ for, otherwise $\delta > 0$ and $\varepsilon_n \to 0$ would exist such that $m(E_{\varepsilon_n}) \leq \tilde{T} - \delta$. In other words, a sequence of measurable subsets A_n of $[0,\tilde{T}]$ would exist such that $m(A_n) \geq \delta$ and

$$|y^{\varepsilon_n}(t)| \leq 2^{1/2}\varepsilon_n \text{ for all } t \in A_n.$$

Since $y^{\varepsilon_n} \to y^*$ uniformly on $[0,\tilde{T}]$ this would imply that

$$|y^*(t)| \leq 2^{1/2}\varepsilon_n + \nu_n \text{ for } t \in A_n$$

where $\nu_n \to 0$. On the other hand, since $y^* \neq 0$ on $[0,\tilde{T}]$, it follows that

$$\lim_{n\to\infty} m\{t; |y^*(t)| \leq 2^{1/2}\varepsilon_n + \nu_n\} = 0.$$

The contradiction we arrived at proves that indeed $\lim_{\varepsilon\to 0} m(E_\varepsilon) = \tilde{T}$. Along with (5.38) this implies the desired equality $\tilde{T} = T^*$. Hence u^* is a time optimal control and the proof is complete.

Letting ε tend to zero in Eq. (5.25) we may regard formally T as the solution to the dynamic programming equation

$$|\partial T(y)| + (Ly, \partial T(y)) = 1 \text{ for all } y \neq 0 \tag{5.39}$$

associated with time optimal control Problem (5.17). We do not know whether T (which is a locally Lipschitz function on T) is indeed a solution to Eq. (5.39). However, by virtue of Proposition 5 we may regard ϕ^ε as an approximating solution of (5.39) and (see 5.29) and (5.28))

$$u^\varepsilon = -\operatorname{sgn} \partial\phi^\varepsilon(y) - \varepsilon\partial\phi^\varepsilon(y), \; y \in H \tag{5.39'}$$

as an approximating feedback control for Problem (5.17). The above procedure for constructing suboptimal feedback control can be extended to certain free time optimal control problems.

As an example consider the following problem; minimize

$$\frac{1}{2}\int_0^T |u(t)|^2 dt + T \tag{5.40}$$

over all $(y,u) \in W^{1,2}_{loc}(\mathbb{R}^+;H) \times L^2_{loc}(\mathbb{R}^+;H)$ subject to

$$\begin{aligned} &y' + Ly = u \quad \text{a.e.} \quad t > 0 \\ &y(0) = y_0, \; y(T) = 0. \end{aligned} \tag{5.41}$$

We associate with (5.40) the infinite time horizon control problem

$$\inf \{\int^\infty (g^\varepsilon(y(t,0,y_0,u)) + \frac{1}{2}|u(t)|^2)dt\} = \psi^\varepsilon(y_0) . \tag{5.42}$$

Then by Proposition 4 there exist $p_\varepsilon \in L^2_{loc}(\mathbb{R}^+;V) \cap L^2(\mathbb{R}^+;H) \cap L^\infty(\mathbb{R}^+;H)$ satisfying the system

$$\begin{aligned} &y'_\varepsilon + Ly_\varepsilon = p_\varepsilon \quad \text{a.e.} \quad t > 0 \\ &p'_\varepsilon + Ap_\varepsilon - p_\varepsilon\partial\beta(y_\varepsilon) = G^\varepsilon(y_\varepsilon), \quad \text{a.e. } t > 0 \\ &p_\varepsilon(t) + \partial\psi^\varepsilon(y_\varepsilon(t)) \ni 0 \text{ for all } t \geq 0 \end{aligned} \tag{5.43}$$

where ψ^ε is the solution to (see Theorem 5)

$$\frac{1}{2}|\partial\psi^\varepsilon(y)|^2 + (Ly, \partial\psi^\varepsilon(y)) = g^\varepsilon(y) \tag{5.44}$$

and

$$u = -\partial\psi^\varepsilon(y) \tag{5.45}$$

is an optimal feedback control for Problem (5.42).

Arguing as in the proof of Proposition 5 we deduce that (see [5]) on some subsequence $\varepsilon \to 0$,

$$u_\varepsilon \to u_1^* \text{ strongly in } L^2(\mathbb{R}^+;H)$$

$$y_\varepsilon \to y_1^* \text{ uniformly on } \mathbb{R}^+ \text{ in } H$$

where (y_1^*,u_1^*) is an optimal pair of Problem (5.40).

Hence formally, we may view ψ (the value function of Problem (5.40)) as solution to the dynamic programming equation (the Bellman equation)

$$\frac{1}{2}|\partial\psi(y)|^2 + (Ly,\partial\psi(y)) = 1 \tag{5.46}$$

corresponding to Problem (5.40).

§6 Galerkin approximation of solutions

Here we shall briefly discuss a Galerkin approximation of Eqs (4.1) and (5.1). We shall suppose that assumptions (i)-(vi) are satisfied and that β is globally Lipschitz.

It will be more convenient to regard A as a linear continuous operator from the space $V = \mathcal{D}((A)^{1/2})$ with the norm $|\cdot|_V$ to its dual $V' \supset H = L^2(\Omega)$. For instance in the special case where A is defined by (4.9), $V = H^1(\Omega)$ and A has an extension from V to V' (again denoted A) given by

$$(Ay,z) = -\sum_{i,j=1}^{N} \int_\Omega a_{ij}(x)y_{x_i}z_{x_j}\,dx$$

$$- \alpha_1\alpha_2^{-1}\int_\Gamma yz\,d\sigma \text{ for all } y,z \in V.$$

By $(\cdot,\cdot)$ we shall denote the pairing between V,V' and the scalar product in H.

Following a well-known scheme (see for instance [1], [13]) we define firstly the internal approximations of spaces V, U and of the operators $L = -A + \beta$ and D.

We are given a parameter h which converges to zero and the following elements:

(a) a finite dimensional Hilbert space V_h

(b) an isomorphism $p_h : V_h \to X_h \subset V$ which is on to X_h

(c) a linear continuous operator $r_h : H \to V_h$

(d) a finite dimensional Hilbert space U_h and an isomorphism $g_h : U_h \to W_h = g_h(U_h)$

(e) a linear continuous operator $s_h . U \to U_h$.

We assume that there exists a constant C independent of h such that

$$|p_h r_h|_{\mathcal{L}(H,V)} + |q_h s_h|_{\mathcal{L}(U,U)} \leq C \text{ for all } h \tag{6.1}$$

and for all $v \in V$ and $u \in U$

$$p_h r_h v \to v \text{ strongly in } V \text{ as } h \to 0 \tag{6.2}$$

$$q_h s_h u \to u \text{ strongly in } U \text{ as } h \to 0. \tag{6.3}$$

In the space V_h one introduces the inner product $(\cdot,\cdot)_h$ and the norms $|\cdot|_h$, $\|\cdot\|_h$,

$$(y_h, z_h)_h = (p_h y_h, p_h z_h) \text{ for all } y_h, z_h \in V_h \tag{6.4}$$

$$|y_h|_h = |p_h y_h| \text{ for all } y_h \in V_h \tag{6.5}$$

$$\|y_h\|_h = |p_h y_h|_h \text{ for all } y_h \in V_h. \tag{6.6}$$

Similarly, in the space U_h introduce the scalar product

$$\langle u_h, v_h \rangle_h = (q_h u_h, q_h v_h)_U \text{ for all } u_h, v_h \in U_h$$

$$|||u_h||| = |q_h u_h|_U \text{ for all } u_h \in U_h.$$

For instance, in the case of the Galerkin method, $V_h = \mathbb{R}^n$, $h = n^{-1}$ with the elements $v_h = (v_h^i;\ 1 \leq i \leq n)$ and $p_h v_h = \sum_{i=1}^{n} v_h^i e_i$, $r_h v = \{(e_i, v)_V;\ i=1,\ldots,n\}$, where $\{e_i\}_{i=1}^{\infty}$ is an orthonormal basis of V (V is a separable Hilbert space). If V is a Sobolev space on Ω (for instance $V = H_0^1(\Omega)$) then we may choose V_h as a space of finite elements on Ω. The dimension of V_h is equal in this case to the number of vertices of triangulation which do not belong to the boundary Γ_h of $\Omega_h = U\Gamma_h$.

For every h define the operator $L_h : V_h \to V_h$ by

$$(L_h y_h, z_h)_h = (L p_h y_h, p_h z_h) \text{ for all } y_h, z_h \text{ in } V_h \tag{6.7}$$

$D_h : V_h \to U_h$ by

$$\langle D_h v_h, u_h \rangle_h = (D p_h v_h, q_h u_h)_U \text{ for all } v_h \in V_h,\ u_h \in U_h \tag{6.8}$$

and the function $g_h : V_h \to \mathbb{R}$, $\Phi_h : U_h \to \mathbb{R}$,

$$g_h(v_h) = g(p_h v_h) \text{ for all } v_h \in V_h \tag{6.9}$$

$$\Phi_h(u_h) = \Phi(q_h u_h) \text{ for all } u_h \in U_h. \tag{6.10}$$

Thus on the space V_h, Eq. (4.1) with final value condition becomes

$$\phi_t^h(t,y_h) - \Phi_h(D_h \phi_y^h(t,y_h)) - (L_h y_h, \phi_y^h(t,y_h))_h$$

$$= -g_h(y_h) \text{ for } t \in [0,T] \text{ and } y_h \in V_h \tag{6.11}$$

$$\phi^h(T,y_h) = 0.$$

This is a classical Hamilton-Jacobi equation on the finite dimensional space V_h. By Theorem 3 Eq. (6.11) has a generalized solution $\phi^h : [0,T] \times V_h \to \mathbb{R}$ given by

$$\phi^h(t,y_h^0) = \inf \{ \int_t^T (g_h(y_h(s)) + \Phi_h^*(u_h(s)))ds \} \tag{6.12}$$

where y_h, u_h satisfy the system

$$y_h'(s) + L_h y_h(s) = -D_h^* u_h(s) \quad \text{a.e.} \quad s \in [t,T]$$

$$y_h(t) = y_h^0 = r_h y_0. \tag{6.13}$$

As remarked earlier, if Φ, β and g are sufficiently smooth then ϕ^h is a classical solution to Eq. (6.11) on a sufficiently small interval $[T-\varepsilon,T]$ and it is unique on this interval. As regards the convergence of the solutions ϕ^h to a solution ϕ to Eq. (4.1) we have the following result:

PROPOSITION 6. *For* $h \to 0$ *we have*

$$\phi^h(t,r_h,y_0) \to \phi(t,y_0) \text{ for all } t \in [0,T],\ y_0 \in V \tag{6.14}$$

where ϕ *is the (generalized) solution to Problem (4.56) given by Eq. (4.43) where* $\psi = 0$.

Proof Since the proof is similar to that of Theorem 5.1 in [2] it will be outlined only. Let $(t,y_0) \in [0,T] \times D(1)$ be fixed and let (y_h^*,u_h^*) be an optimal pair in Problem (6.12). We have

$$\int_t^T (g_h(y_h^*(s)) + \Phi_h^*(u_h^*(s)))ds \leqslant C$$

and by (4.13), (6.10) we see that

$$\int_t^T |||u_h(s)|||_h^2 \, ds \leqslant C_1. \tag{6.15}$$

Now multiplying Eq. (6.13) by y_h^* and integrating over [t,s] we get

$$|y_h^*(s)|_h^2 + \int_t^T \|y_h^*(s)\|_h^2 ds \leqslant C_2 \text{ for all } h . \tag{6.16}$$

Finally, taking the scalar product of (6.13) (in V_h) with y_h^* and integrating over [t,T], we obtain the estimate

$$\int_t^T |y_h^{*\prime}(s)|_h^2 \, ds \leqslant C \text{ for all } h. \tag{6.17}$$

Since by assumptions (iv) (part (4.6)) V is compactly imbedded in H we may infer by (6.15), (6.16), (6.17) that on some sequence $h \to 0$

$$q_h u_h^* \to u_1^* \text{ weakly in } L^2(t,T;U) \tag{6.18}$$

$$p_h y_h^* \to y_1^* \text{ weakly in } L^2(t,T;V) \text{ and strongly in } C([t,T];H) \tag{6.19}$$

$$Lp_h y_h^* \to w \text{ weakly in } L^2(t,T;H). \tag{6.20}$$

This implies by a standard procedure (see for instance [13]) that $w = Ly_1^*$ and

$$\begin{aligned} &(y_1^*)' + Ly_1^* = -D^*u_1^*, \quad \text{a.e. } s \in]t,T[\\ &y_1^*(t) = y_0 \end{aligned} \tag{6.21}$$

Now we shall prove that (y_1^*,u_1^*) is an optimal pair for Problem (4.43), i.e.,

$$\phi(t,y_0) = \int_t^T (g(y_1^*) + \Phi^*(u_1^*))ds. \qquad (6.22)$$

Consider the Cauchy problem

$$z_h' + L_h z_h = -D_h^* s_h u^* \quad \text{a.e.} \quad s \in]t,T[$$

$$z_h(0) = r_h y_0$$

where u^* is an optimal control of Problem (4.43) (where $\psi = 0$). Since $q_h s_h u^* \to u^*$ strongly in $L^2(t,T;U)$ we may conclude that

$$D_h^* s_h u^* \to D^* u^* \text{ strongly in } L^2(t,T;H)$$

and

$$p_h z_h \to y^* \text{ strongly in } C([t,T];H) \text{ and weakly in } L^\infty(t,T;V)$$

$$p_h z_h' \to y^{*\prime} \text{ weakly in } L^2(t,T;H)$$

$$L p_h z_h \to L y^* \text{ weakly in } L^2(t,T;H)$$

where y^* is the solution to Eq. (6.21) corresponding to u^*. Since (y_h^*,u_h^*) is optimal in Problem (6.12) we have

$$\int_t^T (g_h(y_h^*) + \Phi_h^*(u_h^*))ds \leq \int_t^T (g(p_h z_h) + \Phi^*(q_h s_h u^*))ds$$

and therefore

$$\lim_{h\to 0} \phi^h(t,y_h^0) = \int_t^T (g(y_1^*) + \Phi^*(u_1^*))ds = \phi(t,y_0).$$

The latter shows that (y_1^*,u_1^*) is an optimal pair and proves (6.14) as claimed.

By a similar argument it follows that the solution ϕ^h of algebraic equation

$$\Phi_h(D_h \phi_y^h(y_h)) + (L_h y_h, \phi_y^h(y_h))_h = g_h(y_h) \text{ for all } y_h \in V_h \qquad (6.23)$$

given by (see (5.2))

$$\phi^h(y_h^0) = \inf \left\{\int_0^\infty (g_h(y_h(s)) + \Phi_h^*(u_h(s)))ds\right\} \qquad (6.24)$$

where

$$y_h' + L_h y_h = -D_h^* u_h, \quad \text{a.e.} \quad t > 0$$

$$y_h(0) = y_h^0 = r_h y_0$$

approaches the solution ϕ to Eq. (5.1) given by Eq. (5.2). In other words,

$$\lim_{h\to 0} \phi^h(r_h y_0) = \phi(y_0) \text{ for all } y_0 \text{ in } V. \tag{6.25}$$

We notice that in the special case of Problem (5.20), Eq. (6.23) becomes (see Eq. (5.25))

$$\begin{aligned} &\varepsilon/2|\partial\phi^h(y_h)|_h^2 + |\partial\phi^h(y_h)|_h + (L_h y_h, \partial\phi^h(y_h))_h \\ &\quad = g_h^\varepsilon(y_h) \text{ for all } y_h \in V_h. \end{aligned} \tag{6.26}$$

Returning to Eq. (6.11), we recall that by (4.50)

$$u_h = \nabla\Phi_h(D_h\partial\phi^h(t,y_h)), \; t \in [0,T], \quad y_h \in V_h$$

is an optimal feedback controller for Problem (6.12) ($t = 0$). Thus by virtue of Proposition 6 and of relation (6.18) we may regard

$$u = g_h \nabla\Phi_h(D_h\partial\phi^h(t,r_h y)); \; t \in [0,T], \quad y \in V \tag{6.27}$$

as an approximating optimal feedback control of Problem (P).

Similarly, the feedback law (see (5.39)')

$$u = -q_h \operatorname{sgn} \partial\phi^h(r_h y) - \varepsilon q_h \partial\phi^h(r_h, y), \; y \in V \tag{6.28}$$

is an approximating optimal feedback control for the time optimal control Problem (5.17).

The implementation of these suboptimal feedback controllers as well as the calculation of solutions to Eqs (4.1) and (5.1) require numerical methods for Eqs (6.11) and (6.23).

§7 References

[1] J.P. Aubin, *Approximation of elliptic boundary value problems*, Wiley-Interscience, New York, 1972.

[2] V. Arnautu, Approximation of optimal distributed control problems governed by variational inequalities, *Numer. Math.* 38 (1982), 393-416.

[3] V. Barbu, Necessary conditions for distributed control problems governed by parabolic variational inequalities, *SIAM J. Control and Optimiz.* 19 (1981), 64-86.

[4] V. Barbu, Optimal feedback controls for a class of nonlinear distributed parameter systems, *SIAM J. Control and Optimiz.* (To appear)

[5] V. Barbu, Optimal feedback controllers for semilinear parabolic equations, in *Mathematical theory of optimization*, Lecture Notes in Mathematics, Springer, Wiesbaden, 1983.

[6] V. Barbu and G. Da Prato, Hamilton-Jacobi equations and synthesis of nonlinear control processes in Hilbert spaces, *J. Diff. Equations* (To appear)

[7] V. Barbu and G. Da Prato, Local existence for a nonlinear operator equation arising in synthesis of optimal control, *Numer. Funct. Anal. Optimiz.* 1 (6), 1979, 665-667.

[8] V. Barbu, G. Da Prato and C. Popa, Existence and uniqueness for the dynamic programming equation in Hilbert space, *J. Nonlinear Analysis* (To appear)

[9] G. Di Blasio, Synthesis of optimal bilinear controls *J. Math. Anal. and Appl. 88*, 1, 1982, 143-156.

[10] M.C. Delfour and S.K. Mitter, Hereditary differential systems with constant delays, *J. Differential Equations* 12, 1979, 213-235.

[11] I. Ekeland, On the variational principle, *J. Math. Anal. Appl.* 47 (1974) 324-353.

[12] H.O. Fattorini, The time optimal control problem in Banach space, *Appl. Math. Optimiz.* 1 (1974), 163-188.

[13] R. Glowinski, J.L. Lions and R. Tremolières, *Analyse numerique des inéquations variationelles*, Dunod, Paris 1976.

[14] J.L. Lions, *Optimal control of systems governed by partial differential equations*, Springer, Wiesbaden, 1971.

4 The dynamic programming equation for stochastic optimal control

§1 Introduction

In this chapter we shall discuss the Hamilton-Jacobi equations

$$\phi_t(t,x) + \frac{1}{2}|\phi_x(t,x)|^2 - (Ax,\phi_x(t,x)) - \varepsilon/2\ \mathrm{tr}(S\phi_{xx}(t,x))$$
$$= g(t,x);\ t \in [0,T],\ x \in D(A),\quad \varepsilon > 0 \tag{1.1}$$

with the initial value condition:

$$\phi(0,x) = \phi_0(x),\quad x \in H \tag{1.2}$$

in a class of convex functions on a real Hilbert space H which verify suitable growth conditions at infinity. We use here the notations of Chapter 2 (see Introduction); moreover, S is a positive symmetric nuclear operator and for each linear operator U in H the symbol tr(U) denotes the trace of U.

Equation (1.1) is connected with the optimal control problem: minimize

$$E\{\int_0^T (V(t,y(t)) + \frac{1}{2}|u(t)|^2)dt + \phi_0(y(T))\} \tag{1.3}$$

over all u in $M_W^2(0,T;H)$, subject to the differential stochastic equation:

$$dy = (Ay + u)dt + \sqrt{\varepsilon}\ dW_t$$
$$y(0) = x. \tag{1.4}$$

Here, W_t is a H-valued Brownian motion in a probability space (Ω, E, P) such that $\mathrm{cov}(W_t) = tS$, E denotes the expectation, and $V(t,x) = g(T-t,x)$.

Equation (1.1) is extensively discussed by W. Fleming and R. W. Rishel ([9]) in the case where H is finite dimensional and S is nondegenerate. In this case, the existence and uniqueness of a global solution is proved by means of the theory of quasi-linear parabolic equations; then Problem (1.3) is solved using a verification theorem.

In the past few years the case where S is degenerate (and H is finite-dimensional) has been also studied, see P. L. Lions [12].

In this chapter we study Eq. (1.1) as a general Hilbert space and we allow the covariance operator S to be zero in an arbitrary subspace of H.

Our method consists in first approximating Eq. (1.1) by

$$\phi_t^\alpha + \frac{1}{\alpha}(\phi^\alpha - S(\alpha)\phi^\alpha) - (Ax,\phi_x^\alpha) - \varepsilon/2 \ \mathrm{tr}(S\phi_{xx}^\alpha) = 0 \tag{1.5}$$

where $S(\alpha)\phi^\alpha$ is the convex approximation of ϕ^α, and then in solving (1.5) by the stochastic characteristics method. In this way we are also able to show that, as ε goes to zero, the solution ϕ of Eq. (1.1) converges to the linear solution of the Hamilton-Jacobi equation:

$$\begin{aligned} &\phi_t(t,x) + \frac{1}{2}|\phi_x(t,x)|^2 - (Ax,\phi_x(t,x)) = g(t,x) \\ &\phi(0,x) = \phi_0(x) \end{aligned} \tag{1.6}$$

studied in Chapter 2.

Section 2 is devoted to notations. In Section 3 we study the linear problem obtained by the linearization of Eq. (1.1)

$$\begin{aligned} &\phi_t(t,x) - (Ax,\phi_x(t,x)) - \varepsilon/2 \ \mathrm{tr}(S\phi_{xx}(t,x)) = g(t,x) \\ &\phi(0,x) = \phi_0(x) \end{aligned} \tag{1.7}$$

by the characteristics, that is the solution to the stochastic differential equation

$$dx = Axds + \varepsilon dW_s, \ x(t) = x_0, \ s \in [0,T]. \tag{1.8}$$

In Section 4 we study Eq. 1.1 and we use these results to solve Problem (1.3) in Section 5. Finally in Section 6 we study the special case of linear quadratic control by methods similar to those in Chapter 2. This case has been extensively studied, see for example the papers of Ichikawa [11] and Curtain [5], for many results and references.

No attempt has been made in this chapter to study a variational approach of Problem (1.3); we refer to Bensoussan [1] and Davis [8] for an extensive bibliography on this topic. The essential parts of the results of this chapter rely on the authors joint papers [3], [4].

§2 Notations and assumptions

We shall assume throughout that:

(i) H *is a real Hilbert space with inner product* $(\cdot,\cdot)$ *and norm* $|\cdot|$.

(ii) *The linear operator* A *is the infinitesimal generator of a* C_0*-semigroup contraction* e^{tA} *on* H.

(iii) *The linear operator* S *is continuous, nuclear, symmetric, and non negative.*

For any $h = 0,1,\dots$, $n = 0,1,\dots$ and $\phi \in C^h(H)$ we set

$$|\phi|_{h,n} = \sup\{|\phi^{(h)}(x)(1+|x|^{2n})^{-1}|,\ x \in H\} \tag{2.1}$$

$$\|\phi\|_{h,n} = \sup\{|\phi^{(h)}(x) - \phi^{(h)}(y)|\ |x-y|^{-1} (1 + \max(|x|,\ |y|)^{2n})^{-1};\ x,y \in H,\ x \neq y\} \tag{2.2}$$

where $\phi^{(h)}$ stands for the derivative of order h.

It is easy to see that

$$\|\phi\|_{k,n} = |\phi|_{k+1,n},\quad k = 0,1,\dots,\ n = 0,1,\dots \tag{2.3}$$

for any $\phi \in C^{k+1}(H)$.

We shall use the following Banach spaces

$$X_0 = \{\phi \in C(H);\ |\phi|_{X_0} = |\phi|_{0,n_0} < +\infty\} \tag{2.4}$$

$$X_1 = \{\phi \in C^1(H);\ |\phi|_{X_1} = |\phi|_{0,n_0} + |\phi|_{1,n_1} < +\infty\} \tag{2.5}$$

$$X_2 = \{\phi \in C^2(H);\ |\phi|_{X_2} = |\phi|_{0,n_0} + |\phi|_{1,n_1} + |\phi|_{2,n_2} + \|\phi\|_{3,n_3} < +\infty\} \tag{2.6}$$

where $n_0 > n_1 > n_2 > n_3 > 0$ are fixed integers, which we shall choose suitably below. We now set

$$B([0,T];X_i) = \{\phi \in B([0,T];C^i(H));\ \sup_{t\in[0,T]} |\phi(t,\cdot)|_{X_i} < +\infty\}\quad i = 0,1,2. \tag{2.7}$$

Finally, let W be a H-valued Brownian motion in a complete probability space $(\Omega, \mathcal{E}, P)$ such that

$$\mathrm{cov}(W_t) = tS \quad \forall t \geq 0 \tag{2.8}$$

It is well known (see for instance [6]) that there exists an orthonormal basis $\{e_i\}$ in H, a sequence $\{\lambda_i\}$ of non negative real numbers and a sequence $\{\beta_i\}$ of scalar Brownian motions mutually independent, such that:

$$W(t) = \sum_{i=1}^{\infty} \lambda_i \beta_i(t) e_i \tag{2.9}$$

$$Se_i = \lambda_i e_i \quad i = 1,2,... \tag{2.10}$$

We shall denote by $M_W^2(0,T;H)$ the Hilbert space of all measurable H-valued processes $X : [0,T] \times \Omega \to H$ adapted to W and such that:

$$E \int_0^T |X(s)|^2 \, ds < + \infty \tag{2.11}$$

We are going to study the problem

$$\begin{aligned} &\phi_t(t,x) + \tfrac{1}{2}|\phi_x(t,x)|^2 - (Ax, \phi_x(t,x)) \\ &\qquad - \varepsilon/2 \ \mathrm{tr}(S\phi_{xx}(t,x)) = g(t,x) \\ &\phi(0,x) = \phi_0(x) \end{aligned} \tag{2.12}$$

where $\varepsilon > 0$, $\phi_0 \in X_i$ and $g \in B([0,T];X_i)$.

We shall consider first a linearized version of Problem (2.12), namely

$$\begin{aligned} &\phi_t(t,x) - (Ax, \phi_x(t,x)) - \varepsilon/2 \ \mathrm{tr}(S\phi_{xx}(t,x)) = g(t,x) \\ &\phi(0,x) = \phi_0(x) \end{aligned} \tag{2.13}$$

We say that ϕ is a *strong solution* to Problem (2.12) (resp. (2.13)) if ϕ belongs to $B([0,T];X_2) \cap C^1([0,T] \times D(A);\mathbb{R})$ and Eq. (2.12) (resp. (2.13)) is satisfied for all $t \in [0,T]$ and $x \in D(A)$.

Following the definition of Chapter 2, the function $\phi \in B([0,T];X_0)$ is called a *weak solution* to Problem (2.12) if there exist sequences $\{\phi_0^n\} \subset X_2$, $\{g^n\} \subset B([0,T];X_2)$ such that:

$$\phi_0^n \to \phi_0 \text{ in } X_0$$

$$g^n \to g \text{ in } B([0,T];X_0) \tag{2.14}$$

$$\phi^n \to \phi \text{ in } B([0,T];X_0)$$

and ϕ^n are strong solutions to the problems:

$$\phi_t^n(t,x) + \frac{1}{2}|\phi_x^n(t,x)|^2 - (Ax,\phi_x^n(t,x))$$

$$- \quad \varepsilon/2 \ \mathrm{tr}(S\phi_{xx}^n(t,x)) = g^n(t,x) \tag{2.15}$$

$$\phi^n(0,x) = \phi_0(x).$$

§3 The linearized problem

We study first the homogeneous problem

$$\begin{aligned} &\phi_t(t,x) - (Ax,\phi_x^n(t,x)) - \varepsilon/2 \ \mathrm{tr}(S\phi_{xx}(t,x)) = 0 \\ &\phi(0,x) = \phi_0(x) \end{aligned} \tag{3.1}$$

where $\phi_0 \in X_2$.

We shall also consider the approximating problem:

$$\begin{aligned} &\phi_t^n(t,x) - (A_n x,\phi_x^n(t,x)) - \varepsilon/2 \ \mathrm{tr}(S\phi_{xx}^n(t,x)) = 0 \\ &\phi^n(0,x) = \phi_0(x) \end{aligned} \tag{3.2}$$

where $A_n = n^2(n-A)^{-1} - n = nA(n-A)^{-1}$ is the Yosida approximation of A.

LEMMA 1 *Under the hypotheses* (i)-(iii), *if* $\phi_0 \in X_2$, *then the Problem (3.1) has a unique strong solution* $\phi^n \in C^1([0,T] \times H;\mathbb{R}) \cap B([0,T];X_2)$ *given by the formula:*

$$\phi^n(t,x) = E\phi_0(\exp(tA_n)x + \sqrt{\varepsilon} \int_0^t \exp(sA_n)dW_{T-s}) \tag{3.3}$$

Proof Let ϕ^n be defined by formula (3.3); then we have

$$\phi^n \in C^1([0,T] \times H;\mathbb{R}) \cap C([0,T];X_2).$$

Moreover, arguing as in the finite dimensional case (see for instance [10]), it is not difficult to verify that ϕ^n is a solution to Problem (3.3) (see [4] for details).

For the uniqueness assume now that $\psi \in C^1([0,T] \times H;\mathbb{R}) \cap B([0,T];X_2)$ is a strong solution to Problem (3.2) with $\phi_0 = 0$. (If $\psi \in C([0,T];X_2)$ the proof is as in the finite dimensional case.)

Set $\eta(t,x) = \psi(T-t,x)$; then η is a solution to the backward problem

$$\begin{aligned} &\eta_t(t,x) + (Ax,\eta_x(t,x)) + \varepsilon/2 \operatorname{tr}(S\eta_{xx}(t,x)) = 0 \\ &\eta(T,x) = 0. \end{aligned} \tag{3.4}$$

Let $u = u(s) = u(s,t,x)$ be the solution to the stochastic differential equation:

$$du = A_n u ds + \sqrt{\varepsilon}\, dW_s;\ u(t) = x \tag{3.5}$$

i.e.,

$$u(s,t,x) = \exp((s-t)A_n)x + \sqrt{\varepsilon} \int_s^t \exp(s-\sigma)A_n)dW_\sigma. \tag{3.6}$$

We remark now that the second term in formula (3.6) is a martingale. Thus using a well-known martingale inequality (see for instance [14]) we prove that

$$u \in L^p([s,t] \times \Omega;H) \text{ for any } p > 1 \tag{3.7}$$

Set $\eta^k(t,x) = \eta(t,P_k x)$, where

$$P_k x = \sum_{i=1}^k (x,e_i)e_i. \tag{3.8}$$

Clearly $\eta^k \in C^1([0,T] \times H;\mathbb{R}) \cap C([0,T];X_2)$, so we may apply the Itô formula to $\eta^k(s,u)$ (see [14] or [11]); we have

$$\begin{aligned} \eta^k(T,u(T)) &= \eta^k(t,u(t)) + \int_t^T (\eta_s^k(s,u(s)) \\ &+ \varepsilon/2 \operatorname{tr}(S\eta_{xx}^k(s,u(s)))ds + \int_t^T \eta_x^k(s,u(s))dW_s. \end{aligned} \tag{3.9}$$

Taking the expectation we get

$$E(\eta^k(T,u(T)) = \eta^k(t,x)$$

$$+ E\int_t^T (\eta_s^k(s,u(s)) + \varepsilon/2\ tr(S\eta_{xx}^k(s,u(s)))ds. \tag{3.10}$$

In fact, due to (3.7) and the polynomial growth of η^k, we have $\eta_x^k(s,u) \in M_W^2(0,T;H)$. Thus the expectation of the last term in (3.9) vanishes. As k goes to infinity we get $\eta(t,x) = 0$ and the proof of Lemma 1 is complete.

The following corollary follows via a standard variation of constants formula.

COROLLARY 1 *Under the hypotheses (i)-(iii) if* $\phi_0 \in X_2$ *and* $g \in B([0,T];X_2)$, *then Problem (2.13) (with* $A = A_n$*) has a unique strong solution* $\phi^n \in C^1([0,T] \times H;\mathbb{R}) \cap B([0,T];X_2)$ *given by:*

$$\phi^n(t,x) = E\phi_0(\exp(tA_n)x + \sqrt{\varepsilon}\int_0^t \exp(sA_n)dW_{T-s})$$

$$+ E\int_0^t g(s,\exp((t-s)A_n)x + \sqrt{\varepsilon}\int_0^{t-s}\exp(\sigma A_n)dW_{T-\sigma})ds \tag{3.11}$$

PROPOSITION 1 *Under the hypotheses (i)-(iii) if* $\phi_0 \in X_2$, *Problem (3.1) has a unique solution* $\phi \in B([0,T];X_2)$, *given by:*

$$\phi(t,x) = E\phi_0(e^{tA}x + \sqrt{\varepsilon}\int_0^t e^{sA}dW_{T-s}) \tag{3.12}$$

Moreover, for each non negative integer m, *there exists* $\omega_m > 0$ *such that:*

$$|\phi(t,\cdot)|_{i,m} \leq \exp(\varepsilon\omega_m t)|\phi_0|_{i,m};\ i = 0,1,\ldots \tag{3.13}$$

$$||\phi(t,\cdot)||_{i,m} \leq \exp(\varepsilon\omega_{m+1}t)||\phi_0||_{i,m};\ i = 0,1,\ldots \tag{3.14}$$

Proof Let ϕ^n be given by formula (3.3).

As $\exp(tA_n)x \to \exp(tA)x$ uniformly on [0,T] for each $x \in H$, we see that:

$$\phi^n(t,x) \to \phi(t,x)$$

$$tr(S\phi_{xx}^n(t,x)) \to tr(S\phi_{xx}(t,x)) \tag{3.15}$$

for any $x \in H$, uniformly on [0,T].

Moreover, since for all $x \in D(A)$

$$(A_n x, \phi_x^n(t,x)) \to (Ax, \phi_x(t,x)), \text{ uniformly on } [0,T] \tag{3.16}$$

we infer that $\phi_t^n(t,x) \to \phi_t(t,x)$ uniformly on $[0,T]$ and therefore ϕ satisfies Eq. (3.1).

For uniqueness, let η be a solution to Problem (3.1) with $\phi_0 = 0$. For $x \in H$ we set $x_n = n(n-A)^{-1}x$ and consider the equation:

$$\begin{aligned} &\eta_t(t,x_n) - (A_n x_n, \eta_x(t,x_n)) - \varepsilon/2 \ \mathrm{tr}(S\eta_{xx}(t,x_n)) \\ &\qquad = ((A - A_n)x_n, \eta_x(t,x_n)) \end{aligned} \tag{3.17}$$

then, by Corollary 1,

$$\begin{aligned} \eta(t,x_n) = E\int_0^t &((A - A_n)x_n, \eta_x(\exp((t-s)A_n)x_n \\ &+ \sqrt{\varepsilon}\int_0^{t-s} \exp(\sigma A_n) dW_{T-\sigma}) ds. \end{aligned} \tag{3.18}$$

Letting n tend to infinity we get $\eta = 0$ as claimed.

To prove (3.13), we shall restrict ourselves to the case $i = 0$, the other cases being similar. Notice moreover that, due to the (2.3), inequality (3.14) is an immediate consequence of (3.13).

We remark first that, since $\int_0^t e^{sA}\, dW_{T-s}$ is a martingale, it follows (see for instance [14]), that for each $m \in \mathbb{N}$ there exists $\gamma_m > 0$ such that:

$$E\, \Big|\int_0^t e^{sA} dW_{T-s}\Big|^{2m} \leq \gamma_m t^m . \tag{3.19}$$

We have

$$\begin{aligned} |\phi(t,x)| &\leq |\phi_0|_{0,m}\, E(1 + |e^{tA}x + \sqrt{\varepsilon}\int_0^t e^{sA} dW_{T-s}|^{2m}) \\ &= |\phi_0|_{0,m}\, E(1 + (|x|^2 - 2(e^{tA}x + \sqrt{\varepsilon}\int_0^t e^{sA} dW_{T-s} + |\int_0^t e^{sA} dW_{T-s}|^2))^m \\ &\leq |\phi_0|_{0,m}\, E(1 + (|x|^2 - 2(e^{tA}x + \sqrt{\varepsilon}\int_0^t e^{sA} dW_{T-s}) \\ &\quad + |\int_0^t e^{sA} dW_{T-s}|^2)(|x| + |\int_0^t e^{sA} dW_{T-s}|)^{m-1}). \end{aligned} \tag{3.20}$$

Thus, we can conclude that there exist real bounded functions α_p, p = 2,3,...,2m such that:

$$|\phi(t,x)| \leq |\phi_0|_{0,m} E(1 + \sum_{p=2}^{2m} \alpha_p(|x|) |\int_0^t e^{sA} dW_{T-s}|^2)(1+|x|^{2m}). \quad (3.21)$$

Recalling (3.19) we deduce that:

$$\begin{aligned} |\phi(t,x)| &\leq |\phi_0|_{0,m} (1 + |x|^{2m})(1 + \sum_{p=2}^{2m} \alpha_p(\varepsilon t)^{p/2}) \\ &\leq |\phi_0|_{0,m} \exp(\omega_m \varepsilon t) \end{aligned} \quad (3.22)$$

which completes the proof of Proposition 1.

The following corollary is straightforward:

COROLLARY 2 *Under the hypotheses (i)-(iii), if* $\phi_0 \in X_2$ *and* $g \in B([0,T]:X_2)$, *Problem (2.13) has a unique strong solution given by the formula*

$$\begin{aligned} \phi(t,x) &= E\phi_0(e^{tA}x + \sqrt{\varepsilon} \int_0^t e^{sA} dW_{T-s}) \\ &+ E \int_0^t g(s, e^{(t,s)A}x + \sqrt{\varepsilon} \int_0^{t-s} e^{\sigma A} dW_{T-\sigma}) ds. \end{aligned} \quad (3.23)$$

§4 Existence results

We shall study here the Cauchy problem

$$\begin{aligned} &\phi_t(t,x) + \frac{1}{2}|\phi_x(t,x)|^2 - (Ax, \phi_x(t,x)) \\ &\quad - \varepsilon/2 \operatorname{tr}(S\phi_{xx}(t,x)) = g(t,x) \\ &\phi(0,x) = \phi_0(x) \end{aligned} \quad (4.1)$$

where $\phi_0 \in X_2 \cap K$ and $g \in B([0,T];X_2) \cap K$.

Here K denotes the cone of all convex mappings $H \to \mathbb{R}$ and $\mathcal{K} = \{\phi \in C([0,T] \times H;\mathbb{R}); \phi(t,\cdot) \in K \;\; \forall t \in [0,T]\}$.

As in Chapter 2 (§4) we consider the approximating problem

$$\phi_t^\alpha(t,x) + \frac{1}{\alpha}(\phi^\alpha(t,x) - \phi_\alpha^\alpha(t,x)) - (Ax,\phi_x^\alpha(t,x))$$

$$- \varepsilon/2 \ \mathrm{tr}(S\phi_{xx}^\alpha(t,x)) = g(t,x), \ t \in [0,T], \ x \in D(A) \tag{4.2}$$

$$\phi^\alpha(0,x) = \phi_0(x)$$

and the following weak form of (4.2)

$$\begin{aligned} \phi^\alpha(t,x) &= e^{-t/\alpha} \, E\phi_0(e^{tA}x + \sqrt{\varepsilon}\int_0^t e^{\sigma A}dW_{T-s}) \\ &+ E\int_0^t e^{-(t-s)/\alpha}(\frac{1}{\alpha}\,\phi_\alpha^\alpha + g)(s,e^{(t-s)A}x + \sqrt{\varepsilon}\int_0^{t-s} e^{\sigma A}dW_{T-\sigma})ds. \end{aligned} \tag{4.3}$$

Here ϕ_α is the convex regularization of ϕ.

We shall prove existence of a solution to (4.1) under hypotheses (i)-(iii) and moreover

(iv) $n_0 \geq 2n_1(1 + n_2)$.

LEMMA 2 *Assume that* $\phi,\bar{\phi} \in X_2 \cap K$. *Then for every* $n = 0,1,\ldots$ *there exists a continuous positive function* C_n *such that*

$$|\phi_\alpha|_{0,n} \leq (1 + \alpha C_n(|F(0)|)) \ |\phi|_{0,n} \tag{4.4}$$

$$|\phi_\alpha|_{1,n} \leq |\phi|_{1,n} \tag{4.5}$$

$$|\phi_\alpha|_{2,n} \leq |\phi|_{2,n} \tag{4.6}$$

$$\|\phi_\alpha\|_{2,n} \leq (1 + \alpha C_n(|F(0)|)) \, \|\phi\|_{2,n} \tag{4.7}$$

$$|\phi_\alpha - \bar{\phi}_\alpha|_{0,n} \leq (1 + \alpha C_n(|F(0)|)) \, |\phi - \bar{\phi}|_{0,n} \tag{4.8}$$

$$|\phi_\alpha - \bar{\phi}_\alpha|_{1,n} \leq (1 + \alpha C_n(|F(0)|)) \, |\phi - \bar{\phi}|_{1,n} \tag{4.9}$$

where $F = \phi_x$.

Moreover, if $|\phi|_{X_2}$, $|\bar{\phi}|_{X_2} \leq \lambda$ *then there exists* $c(\lambda) > 0$ *such that*

$$|\phi_\alpha - \bar{\phi}_\alpha|_{0,n_0} + |\phi_\alpha - \bar{\phi}_\alpha|_{1,n_1} + |\phi_\alpha - \bar{\phi}_\alpha|_{2,n_1+n_2}$$
$$\leq (1 + \alpha c(\lambda))\{|\phi - \bar{\phi}|_{0,n_0} + |\phi - \bar{\phi}|_{1,n_1} + |\phi - \bar{\phi}|_{2,n_1+n_3}\} \tag{4.10}$$

Proof Set $x_\alpha = (1 + \alpha F)^{-1}(x)$, then $|x_\alpha| \leq |x| + \alpha\,|F(0)|$ and there exists a continuous positive function C_n such that

$$1 + |x_\alpha|^{2n} \leq (1 + \alpha C_n(|F(0)|)(1 + |x|^{2n}). \tag{4.11}$$

Now all the inequalities are a simple (although tedious) consequence of properties of ϕ_α (see Chapter 2).

LEMMA 3 *Assume that hypotheses (i)-(iii) hold. Then if* $\phi_0 \in X_2 \cap K$ *and* $g \in B([0,T];X_2) \cap K$ *Eq. (4.3) has a unique solution* $\phi^\alpha \in B([0,T];X_2)$. *Moreover, there exists* $\tilde{\omega}_i \geq 0$, $i = 0,1,2,3$ *such that*

$$|\phi^\alpha(t,\cdot)|_{i,n_i} \leq \exp(\varepsilon\tilde{\omega}_i t)|\phi_0|_{i,n_i}$$
$$+ \int_0^t \exp\,(\varepsilon\tilde{\omega}_i(t-s))|g(s,\cdot)|_{i,n_i}ds,\ i = 0,1,2, \tag{4.12}$$

$$\|\phi^\alpha(t,\cdot)\|_{2,n_3} \leq \exp\,(\varepsilon\tilde{\omega}_3 t)\|\phi_0\|_{2,n_3}$$
$$+ \int_0^t \exp\,(\varepsilon\tilde{\omega}_3(t-s))\|g(s,\cdot)\|_{2,n_3}ds. \tag{4.13}$$

Finally, ϕ^α *is the unique strong solution to (4.2).*

Proof Set

$$\phi^0(t,x) = e^{-t/\alpha}E\phi_0(e^{tA}x + \sqrt{\varepsilon}\int_0^t e^{sA}dW_{T-s})$$
$$+ E\int_0^t e^{-(t-s)/\alpha}g(s,e^{(t-s)A}x + \sqrt{\varepsilon}\int_0^{t-s} e^{\sigma A}dW_{T-\sigma})ds$$
$$\phi^{n+1}(t,x) = \phi^0(t,x) + \frac{1}{\alpha}E\int_0^{t-s} e^{-(t-s)/\alpha} \tag{4.14}$$
$$\phi_\alpha^n(s,e^{(t-s)A}x + \sqrt{\varepsilon}\int_0^{t-s} e^{\sigma A}dW_{T-\sigma})ds.$$

By use of inequalities (4.4)-(4.9) and (3.13)-(3.14) it is not difficult to find $\tilde{\omega}_i > 0$, $i = 0,1,2,3$ such that

$$|\phi^n(t,\cdot)|_{i,n_i} \leqslant \exp(\varepsilon\tilde{\omega}_i t)|\phi_0|_{i,n_i}$$

$$+ \int_0^t \exp(\varepsilon\tilde{\omega}_i(t-s))|g(s,\cdot)|_{i,n_i} ds \quad i = 0,1,2 \tag{4.15}$$

$$||\phi^n(t,\cdot)||_{2,n_3} \leqslant \exp(\varepsilon\tilde{\omega}_3 t)||\phi_0||_{2,n_3}$$

$$+ \int_0^t \exp(\varepsilon\tilde{\omega}_3(t-s))||g(s,\cdot)||_{2,n_3} ds. \tag{4.16}$$

By (4.15) and (4.16) we see that the set

$$\Gamma = \{\phi^n(t,\cdot);\ n \in \mathbb{N},\ t \in [0,T]\}$$

is bounded in X_2. Set now

$$|\phi|_\Sigma = |\phi|_{0,n_0} + |\phi|_{1,n_1} + |\phi|_{2,n_1+n_3}. \tag{4.17}$$

By (4.13) it follows (Γ being bounded in X_2) that $\{\phi^n(t,\cdot)\}$ is a Cauchy sequence with respect to the norm $|\ |_\Sigma$, uniformly in t. This implies that $\{\phi^n\}$ converges in $B([0,T];C^2(H))$ to a function ϕ^α. By (4.15) and (4.16) it follows that $\phi^\alpha \in B([0,T];X_2)$ and also that (4.12) and (4.13) hold.

Finally the last claim is a simple verification.

LEMMA 4 *Assume that* $\phi \in C^2(H) \cap K$ *and let* $n_0 \geqslant n_1 \geqslant n_2 \geqslant 0$ *non negative integers. Let* $R_{\phi,\alpha}$ *defined by Eq.(4.15) in Chapter 2. Then, there exists an increasing mapping* $\gamma : \mathbb{R}^2 \to \mathbb{R}_+$ *such that*

$$|R_{\phi,\alpha}|_{0,n_0} \leqslant \alpha\gamma(|\phi|_{1,n_1}, |\phi|_{2,n_2}). \tag{4.18}$$

If moreover m_1 *is another integer such that*

$$m_1 \geqslant \max\{2n_1 + 2n_1n_3,\ n_1 + 2n_2\} \tag{4.19}$$

then there exists an increasing mapping $\eta : \mathbb{R}^3 \to \mathbb{R}^+$ *such that*

$$|R_{\phi,\alpha}|_{1,m_1} \leqslant \alpha\eta(|\phi|_{1,n_1}, |\phi|_{2,n_2}, ||\phi||_{2,n_3}). \tag{4.20}$$

Proof It is a straightforward verification which is established by arguing in the same way as in obtaining the inequalities (4.16) and (4.34) of Chapter 2.

The main existence and uniqueness result for Eq. (4.1) is the following:

THEOREM 1 *Assume the hypotheses (i)-(iv) are fulfilled. Let* $\phi_0 \in X_2 \cap K$ *and* $g \in B([0,T];X_2) \cap K$. *Then the Cauchy Problem (4.1) has a unique weak solution* $\phi \in B([0,T];X_1)$ *such that* $\phi(\cdot,x) \in W^{1,\infty}([0,T])$ *for all* $x \in D(A)$. *Moreover, the set* $\{\phi(t,\cdot);t \in [0,T]\}$ *is bounded in* X_2 *and for each* $x \in D(A)$, Eq. (4.1) *is verified for almost every* $t \in [0,T]$. *Finally the map* $(\phi_0,g) \to \phi$ *is Lipschitz from* $X_0 \times B([0,T];X_0)$ *to* $B([0,T];X_0)$.

REMARK 1 From the last claim of Theorem 1 it follows that if X_2 is dense in X_0 (as in the finite dimensional case), then for any $\phi_0 \in X_0$ and $g \in B([0,T];X_0)$ Problem (4.1) has a unique weak solution $\phi \in B([0,T];X_0)$.

Proof of Theorem 1

We shall obtain the solution ϕ to (4.1) as the limit for $\alpha \to 0$ of ϕ^α. The first step is the proof of the convergence of ϕ^α. Let $\alpha, \beta > 0$; by (4.15) of Chapter 2 it follows that

$$\phi_t^\beta + \frac{1}{\alpha}(\phi^\beta - \phi_\alpha^\beta) - (Ax,\phi_x^\beta) - \varepsilon/2 \operatorname{tr}(S\phi_{xx}^\beta)$$

$$= g + R_{\phi^\beta,\alpha} - R_{\phi^\beta,\beta} \tag{4.21}$$

$$\phi^\beta(0,x) = \phi_0(x).$$

By Corollary 2 we get

$$\phi^\beta(t,x) = e^{-t/\alpha}E\phi_0(e^{tA}x + \sqrt{\varepsilon}\int_0^t e^{sA}dW_{T-s})$$

$$+ E\int_0^t e^{-(t-s)/\alpha}(\frac{1}{\alpha}\phi_\alpha^\beta + R_{\phi^\beta,\alpha} - R_{\phi^\beta,\beta}) \tag{4.22}$$

$$(s,e^{(t-s)A}x + \sqrt{\varepsilon}\int_0^{t-s} e^{\sigma A}dW_{T-\sigma})ds.$$

By (4.12), there exists $\lambda > 0$ such that

$$|\phi^{\alpha}(t,\cdot)|_{X_2} \leq \lambda,\ \forall\alpha > 0,\ \forall t \in [0,T]. \tag{4.23}$$

Moreover by virtue of (3.13), (4.8) and (4.18) we have

$$|\phi^{\beta}(t,\cdot) - \phi^{\alpha}(t,\cdot)|_{0,n_0} \leq \int_0^t \exp((t-s)(\varepsilon\omega_{n_0} - \frac{1}{\alpha}))$$
$$(\frac{1}{\alpha}|\phi^{\beta} - \phi^{\alpha}|_{0,n_0}(1 + C_{n_0}(\lambda)) + (\alpha + \beta)\gamma(\lambda,\lambda))ds. \tag{4.24}$$

By Gronwall's lemma, we see that there exists a constant C such that

$$|\phi^{\beta}(t,\cdot) - \phi^{\alpha}(t,\cdot)|_{0,n_0} \leq C(\alpha + \beta). \tag{4.25}$$

Here, there exists $\phi \in B([0,T];X_0)$ such that $\phi^{\alpha} \to \phi$ in $B([0,T];X_0)$ and, by (4.21), ϕ is a weak solution to Problem (4.1). We shall study now the convergence of ϕ_x^{α}. Choose $m_1 > \max\{2n_1 + 2n_1n_3, n_1 + 2n_2\}$; arguing as in the proof of (4.24), we get

$$|\phi^{\alpha}(t,\cdot) - \phi^{\beta}(t,\cdot)|_{1,m_1} \leq \int_0^t \exp((t-s)(\omega_{m_1} - \frac{1}{\alpha})$$
$$(\frac{1}{\alpha}(1 + \alpha C_{m_1}(\lambda))|\phi^{\alpha}(s,\cdot) - \phi^{\beta}(s,\cdot)|_{1,m_1} \tag{4.26}$$
$$+ (\alpha + \beta)\eta(\lambda,\lambda,\lambda))ds.$$

Using once again Gronwall's lemma we set

$$|\phi^{\alpha}(t,\cdot) - \phi^{\beta}(t,\cdot)|_{1,m_1} \leq C_1(\alpha + \beta) \tag{4.27}$$

and therefore $\phi \in B([0,T];C^1(H))$. By (4.12) it follows that $\phi \in B([0,T];X_1)$.

We now show that $\phi(t,\cdot) \in X_2$. By Lemma 3, for any $R > 0$ there exists $M_R > 0$ such that:

$$|\phi_{xx}^{\alpha}(t,\cdot)|_{0,R} + ||\phi_{xx}^{\alpha}(t,\cdot)||_{0,R} \leq M_R. \tag{4.28}$$

We fix t and set

$$\zeta_{\alpha,x,y,u}(h) = \zeta_{\alpha}(h) = (u,\phi_x^{\alpha}(t,x + hy)), \tag{4.29}$$
$$|x| \leq R,\ h \in [-1,1].$$

It follows that

$$\zeta_\alpha'(h) = \phi_{xx}^\alpha(t,x + hy)(u,y)$$

and, by (4.27)

$$|\zeta_\alpha'(h) - \zeta_\alpha'(k)| \leqslant M_{2R}|h - k|\ |u||y|^2 \quad h,k \in [-1,1].$$

By the Ascoli-Arzelà theorem, there exists a sequence $\alpha_k \to 0$ such that

$$\begin{aligned} \zeta_{\alpha_k}(h) &= (u,\phi_x^{\alpha_k}(t,x + hy)) \to (u,\phi_x(t,x + hy)) \\ \zeta_{\alpha_k}'(h) &\to \frac{d}{dh}(u,\phi_x(t,x + hy)). \end{aligned} \tag{4.30}$$

We set

$$\left[\frac{d}{dh}(u,\phi_x(t,x + hy))\right]_{h=0} = G\phi(t,x)(u,y).$$

From (4.30) it follows that

$$\phi_{xx}^\alpha(t,x)(u,y) \to G\phi(t,x)(u,y). \tag{4.31}$$

To prove that $G\phi = \phi_{xx}$ it suffices to show that G is continuous in x. We have:

$$\begin{aligned} &|\zeta_{\alpha,x,y,u}'(0) - \zeta_{\alpha,z,y,u}'(0)| \\ &= |\phi_{xx}^\alpha(t,x)(u,y) - \phi_{xx}^\alpha(t,z)(u,y)| \leqslant M_{2R}|x - z||u||y| \end{aligned} \tag{4.32}$$

and, recalling (4.31), we get for $k \to \infty$

$$|(G\phi(t,x) - G\phi(t,z))(u,y)| \leqslant M_{2R}|x - z||u||y|. \tag{4.33}$$

It follows that $G\phi(t,x) = \phi_{xx}(t,x)$ and

$$\phi_{xx}^\alpha(t,x)(u,y) \to \phi_{xx}(t,x)(u,y) \tag{4.34}$$

which implies $\phi(t,\cdot) \in X_2$ and moreover

$$\mathrm{tr}(S\phi_{xx}^\alpha(t,x)) \to \mathrm{tr}(S\phi_{xx}(t,x)). \tag{4.35}$$

Recalling now that

$$\phi_t^\alpha + \frac{1}{2}|\phi_x^\alpha|^2 - (Ax,\phi_x^\alpha) - \varepsilon/2\ \mathrm{tr}(S\phi_{xx}^\alpha) = g - R_{\phi^\alpha,\alpha}$$

and keeping in mind Estimate (4.18), we may infer that for $\alpha \to 0$ and $x \in D(A)$

$$\phi_t^\alpha(t,x) \to -\frac{1}{2}|\phi_x(t,x)|^2 + (Ax,\phi_x(t,x))$$

$$+ \varepsilon/2\ \mathrm{tr}(S\phi_{xx}t,x)) + g(t,x).$$

This implies that $\phi(\cdot,x) \in W^{1,\infty}([0,T])$ and, moreover, that ϕ verifies Eq. (4.1) a.e. It remains to prove uniqueness.

Let ϕ^i, $i = 1,2$ be two solutions to Problem (4.1) corresponding to (ϕ_0^i,g^i). We have

$$\phi_t^i + \frac{1}{\alpha}(\phi^i - \phi_\alpha^i) - (Ax,\phi_x^i) - \varepsilon/2\ \mathrm{tr}(S\phi_x^i) = g^i + R_{\phi^i,\alpha}$$

$$\phi^i(t,x) = \phi_0^i.$$

By virtue of Proposition 1, this yields

$$\phi^i(t,x) = e^{-t/\alpha}E\phi_0^i(e^{tA}x + \sqrt{\varepsilon}\int_0^t e^{sA}dW_{T-s})$$

$$+ E\int_0^t e^{-(t-s)/\alpha}(\frac{1}{\alpha}\phi_\alpha^i + g^i)(s,e^{(t-s)A} + \sqrt{\varepsilon}\int_0^{t-s} e^{\sigma A}dW_{T-\sigma})ds.$$

Next, by (3.13), (4.8) and (4.18) it follows via Gronwall's lemma that

$$|\phi^1(t,\cdot) - \phi^2(t,\cdot)|_{0,n_0} \leq C|\phi_0^1 - \phi_0^2|_{0,n_0}$$

$$+ \int_0^t |g^1(s,\cdot) - g^2(s,\cdot)|_{0,n_0} ds.$$

This implies uniqueness and the last claim of Theorem 1.

We shall now study the convergence of $\{\phi^{(\varepsilon)}\}$ for $\varepsilon \to 0$; here we denote by $\phi^{(\varepsilon)}$ the solution to Eq. (4.1).

THEOREM 2 *Under the assumptions of Theorem* 1 $\phi^{(\varepsilon)} \to \phi$ *in* $B([0,T];X_0) \cap B([0,T];C^1(H))$ *where* ϕ *is a strong solution to the Hamilton–Jacobi equation*

$$\phi_t(t,x) + \frac{1}{2}|\phi_x(t,x)|^2 - (Ax,\phi_x(t,x)) = g(t,x)$$

$$\text{a.e.}\quad t \in [0,T],\quad x \in D(A) \tag{4.36}$$

$$\phi(0,x) = \phi_0(x).$$

Proof First we observe that the constants appearing in Estimates (4.12) and (4.13), can be taken to be independent of ε. Next, for $\varepsilon,\lambda, > 0$ we have

$$\phi_t^{(\lambda)} + \frac{1}{2}|\phi_x^{(\lambda)}|^2 = (Ax,\phi_x^{(\lambda)}) - \varepsilon/2\ \mathrm{tr}(S\phi_{xx}^{(\lambda)})$$

$$= g + \frac{\lambda-\varepsilon}{2}\ \mathrm{tr}(S\phi_{xx}^{(\lambda)}).$$

Now, arguing as in the proof of (4.25) and (4.27), we get

$$|\phi^{(\lambda)}(t,\cdot) - \phi^{(\varepsilon)}(t,\cdot)|_{0,n_0} + |\phi^{(\lambda)}(t,\cdot) - \phi^{(\varepsilon)}(t,\cdot)|_{1,m_1}$$

$$\leq C(\varepsilon + \lambda)$$

where m_1 is chosen as in the statement below (4.19).

It follows that $\phi^{(\varepsilon)} \to \phi$ in $B([0,T];X_0) \cap B([0,T];C^1(H))$. Again, by Estimates (4.12) and (4.13), it follows that $\{\phi^{(\varepsilon)}(t,\cdot)\}$ remains in a bounded subset of X_2 and for $\varepsilon \to 0$, $t \in [0,T]$ and $x \in D(A)$ we have

$$\phi_t^{(\varepsilon)}(t,x) \to -\frac{1}{2}|\phi_x(t,x)|^2 + (Ax,\phi_x(t,x)) + g(t,x)$$

uniformly in $[0,T]$. This completes the proof of Theorem 2.

We shall now give another approximation result, which will be useful in the section.

PROPOSITION 2 *Assume that the hypotheses of Theorem 1 hold; for any* $n \in \mathbb{N}$, *let* ϕ^n *be the solution to the following problem:*

$$\phi_t^n + \frac{1}{2}|\phi_x^n|^2 + (A_n x,\phi_x^n) - \varepsilon/2\ \mathrm{tr}(S\phi_{xx}^n) = g$$

$$\phi^n(0,x) = \phi_0(x) \tag{4.37}$$

and let ϕ *be the solution of Problem (4.1). Then* $\phi^n \to \phi$ *in* $B([0,T];X_0) \cap B([0,T];C^1(H))$.

Proof Let $\phi^{n,\alpha}$ be the solution to

$$\begin{aligned} &\phi_t^{n,\alpha} + \frac{1}{\alpha}(\phi^{n,\alpha} - \phi_\alpha^{n,\alpha}) + (A_n x, \phi_x^{n,\alpha}) - \varepsilon/2 \operatorname{tr}(S\phi_{xx}^{n,\alpha}) = g \\ &\phi^{n,\alpha}(0,x) = \phi_0(x). \end{aligned} \tag{4.38}$$

Proceeding as in the proof of Theorem 1, we see that

$$\lim_{n\to\infty} \phi^{n,\alpha} = \phi^\alpha \quad \text{in } B([0,T];X_0) \cap B([0,T];C^1(H))$$

uniformly on α. Hence, for $n \to \infty$

$$|\phi - \phi^n|_{X_0} \leq |\phi - \phi^\alpha|_{X_0} + |\phi^\alpha - \phi^{n,\alpha}|_{X_0} + |\phi^{n,\alpha} - \phi^n|_{X_0} \to 0 \tag{4.39}$$

REMARK 2 $1^{\underline{o}}$ Assume that the hypotheses of Theorem 1 hold and, moreover, that

$$\begin{aligned} &\text{(i)} \quad \phi_0 \in K \cap C^k(H),\ g \in K \cap B([0,T];C^k(H)) \\ &\text{(ii)} \quad |\phi_0|_{i,n_i} < +\infty,\ i = 1,\dots,k,\ \|\phi_0\|_{k,n_{k+1}} \\ &\text{(iii)} \quad \sup_{t\in[0,T]} |g(t,\cdot)|_{i,n_i} < +\infty \quad i = 1,\dots,k \\ &\qquad \sup_{t\in[0,T]} \|g(t,\cdot)\|_{k,n_{k+1}} < +\infty. \end{aligned} \tag{4.40}$$

Then, if $n_0 \geq n_1 \geq \dots \geq n_{k+1}$ are suitably chosen, it is possible to generalize the foregoing arguments and prove that

$$\phi \in B([0,T];C^{k+1}(H))$$

and

$$\sup\{\|\phi(t,\cdot)\|_{k,n_{k+1}},\ t \in [0,T]\} < +\infty.$$

In particular, for $k \geq 3$, ϕ is a strong solution to Problem (4.1). It is also possible to get a classical solution of Eq. (4.1) (that is $\phi \in C^1([0,T] \times H)$) under hypotheses similar to hypothesis (5.1) of Chapter 2.

$2^{\underline{o}}$ Assume $|e^{tA}| \leq e^{\omega t}$ instead of (ii); then, arguing as in Remark 6 of Chapter 2 and setting

$$\chi(t,x) = \phi(t,e^{-\omega t}x)$$

we reduce Eq. (4.1) to

$$\chi_t + b(t)|\chi_x|^2 - (A'x,\chi_x) - \varepsilon/2\ \mathrm{tr}(S\chi_{xx}) = g'(t,x) \tag{4.41}$$

where $\dot{b}(t) = e^{2\omega t}$, $A' = A - \omega, g'(t,x) = g(t,e^{-\omega t}x)$.

3º We can generalize Theorem 1 of Chapter 3 to the nonconvex case for Eq. (4.1), when A is nonlinear thus obtaining local solutions.

§5 Synthesis of optimal control

Consider the optimal control problem: Minimize

$$E\{\int_0^T (g(x(t)) + \frac{1}{2}|u(t)|^2)dt + \phi_0(x(T))\} \tag{5.1}$$

over all $u \in M_W^2(0,T;H)$, subject to

$$\begin{aligned} dx &= (Ax + u)dt + \sqrt{\varepsilon}\, dW_t \\ &\qquad\qquad t \in [0,T] \\ x(0) &= x_0. \end{aligned} \tag{5.2}$$

Here we assume hypotheses (i)-(iv) and moreover

$$\phi_0,\ g \in K \cap X_2. \tag{5.3}$$

By solution to (5.2) we shall mean 'mild' solutions (see Chapter 1). The Bellman equation associated with Problem (5.1) is the following

$$\begin{aligned} &\psi_t(t,x) - \frac{1}{2}|\psi_x(t,x)|^2 + (Ax,\psi_x(t,x)) \\ &\qquad + \varepsilon/2\ \mathrm{tr}(S\psi_{xx}(t,x)) + g(x) = 0 \\ &\psi(T,x) = \phi_0(x). \end{aligned} \tag{5.4}$$

The substitution $\phi(t,x) = \psi(T-t,x)$ reduces Eq. (5.4) to Eq. (4.1). By virtue of Theorem 1, Problem (5.4) has a unique weak solution $\psi \in B([0,T];X_1)$; moreover

$$\sup_{t\in[0,T]} |\psi(t,\cdot)|_{X_2} < +\infty. \tag{5.5}$$

We consider also the approximating problems

$$\psi^n_t(t,x) - \frac{1}{2}|\psi^n_x(t,x)|^2 + (A_n x,\psi^n_x(t,x))$$

$$+ \varepsilon/2\ \mathrm{tr}(S\psi^n_x(t,x)) + g(x) = 0 \tag{5.6}$$

$$\psi^n(T,x) = \phi_0(x)$$

where $A_n = nA(n - A)^{-1}$, and

$$\psi^{n,\alpha}_t(t,x) - \frac{1}{\alpha}(\psi^{n,\alpha}(t,x) - \psi_\alpha{}^{n,\alpha}(t,x)) + (A_n x,\psi^{n,\alpha}_x(t,x))$$

$$+ \varepsilon/2\ \mathrm{tr}(S\psi^{n,\alpha}_{xx}(t,x)) + g(x) = 0 \tag{5.7}$$

$$\psi^{n,\alpha}(T,x) = \phi_0(x).$$

Here $\psi^{n,\alpha}_\alpha$ is the convex regularization of $\psi^{n,\alpha}$.

By Proposition 2, $\psi^n \to \psi$ in $B([0,T];X_0) \cap B([0,T];C^1(H))$ and by Theorem 1 we have

$$\psi^{n,\alpha}_t - \frac{1}{2}|\psi^{n,\alpha}_x|^2 + (A_n x,\psi^{n,\alpha}_x) + \varepsilon/2\ \mathrm{tr}(S\psi^{n,\alpha}_{xx}) \tag{5.8}$$

$$+ g(x) - R_{\psi^{n,\alpha},\alpha} = 0$$

$$\lim_{\alpha\to 0} \psi^{n,\alpha} = \psi^n \text{ in } B([0,T];X_0) \cap B([0,T];C^1(H)) \tag{5.9}$$

and

$$\sup\{|\psi^{n,\alpha}(t,\cdot)|_{X_1};\ t \in [0,T],\ \alpha > 0,\ n \in \mathbb{N}\} < +\infty. \tag{5.10}$$

LEMMA 5 *Assume (i)-(iv) and (5.3). Let* $u \in M^2_W(0,T;H)$, $x_0 \in H$, x *the solution to*

$$dx(s) = (Ax + u)ds + \sqrt{\varepsilon}\, dW_s$$

$$x(t) = x_0 \tag{5.11}$$

and ψ *the solution to (5.4). Then we have:*

$$\psi(t,x_0) + \frac{1}{2} E \int_t^T |\psi_x(s,x(s)) + u(s)|^2 \, ds$$
$$= E\{\int_t^T (g(x(s)) + \frac{1}{2}|u(s)|^2)ds + \phi_0(x(T))\} \tag{5.12}$$

Proof Let x_n be the solution to the following problem

$$dx_n = (A_n x_n + u)ds + \sqrt{\varepsilon}\, dW_s$$
$$x_n(t) = x_0 \tag{5.13}$$

and apply the Itô formula to $\psi^{n,\alpha}(s,x_n)$; this would not be correct because the second derivative $\psi_{xx}^{n,\alpha}$ is only weak continuous. Moreover, we can use the same argument as in the proof of Lemma 1 to show that:

$$E\psi^{n,\alpha}(T,x_n(T)) = \psi^{n,\alpha}(t,x) + E \int_t^T (\psi_s^{n,\alpha}(s,x_n)$$
$$+ (A_n x_n + u, \psi_x^{n,\alpha}(s,x_n)) + \varepsilon/2 \, tr(S\psi_{xx}^{n,\alpha}(s,x_n)))ds \tag{5.14}$$

from which

$$\psi^{n,\alpha}(t,x) + E \int_t^T \frac{1}{2}|\psi_x^{n,\alpha}(s,x_n) + u(s)|^2 \, ds$$
$$= E \{\int_t^T (g(x_n(s) + \frac{1}{2}|u(s)|^2)ds + \phi_0(x_n(T))$$
$$- \int_t^T R_{\psi^{n,\alpha},\alpha}(s,x_n)ds\}. \tag{5.15}$$

As $\alpha \to 0$, recalling (4.18) and (5.10), we get:

$$\psi^n(t,x) + E \int_t^T \frac{1}{2}|\psi_x^n(s,x_n) + u(s)|^2 ds$$
$$= E\{\int_t^T (g(x_n(s)) + \frac{1}{2}|u(s)|^2)ds + \phi_0(x_n(T))\} \tag{5.16}$$

and as $n \to \infty$ the conclusion follows.

We now prove that $\psi(t,x)$ is the optimal value function of Problem (5.1).

PROPOSITION 3 *Assume (i)-(iv) and (5.3). Let* ψ *be the weak solution to Problem (5.4). Then, for all* $t \in [0,T]$ *and* $x_0 \in H$ *we have:*

$$\psi(t,x_0) = \inf \{E(\int_t^T (g(x(s)) + \tfrac{1}{2}|u(s)|^2)ds + \phi_0(x(T)));$$
$$u \in M_W^2(t,T;H),\ dx(s) = (Ax + u)ds + \sqrt{\varepsilon}\, dW_s\} \tag{5.17}$$

Moreover the solution x^* *to the problem:*

$$dx = (Ax - \psi_x(t,x))dt + \sqrt{\varepsilon}\, dW_t$$
$$x(0) = x_0 \tag{5.18}$$

is an optimal trajectory corresponding to the optimal control u^* *given by:*

$$u^*(t) = -\psi_x(t,x^*(t)) \quad \text{a.e. } t \in [0,T]. \tag{5.19}$$

The optimal control u^* *is unique.*

Proof By Lemma 5 we have

$$\psi(t,x_0) \leqslant \inf \{E(\int_t^T (g((s)) + \tfrac{1}{2}|u(s)|^2)ds + \phi_0(x(T)));$$
$$dx(s) = (Ax + u)ds + \sqrt{\varepsilon}\, dW_s,\ t \leqslant s \leqslant T,\ x(0) = x_0, \tag{5.20}$$
$$u \in M_W^2(t,T;H)\}.$$

Now the Cauchy Problem (5.18) has a unique global solution x^* (see Proposition 15 in Chapter 1) because ψ_x is monotone. From (5.12) with $u^*(s) = -\psi_x(s,x^*(s))$ we get

$$\psi(t,x_0) = E \{\int_t^T (g(x^*(s)) + \tfrac{1}{2}|u^*(s)|^2)ds + \phi_0(x^*(T)) \tag{5.21}$$

and therefore u^* is optimal and ψ coincides with the value function of Problem (5.1).

Assume now that $(\tilde{u},\tilde{x})$ is another optimal pair. Again by Formula (5.12) it follows that

$$E \int_t^T |\psi_x(s,\tilde{x}) + \tilde{u}(s)|^2 ds = 0 \tag{5.22}$$

which implies $\tilde{u} = -\psi_x(s,\tilde{x}(s))$. Since the solution to Problem (5.18) is unique, we infer that $\tilde{x} = x^*$ and $\tilde{u} = u^*$, as claimed.

§6 Linear quadratic control

Consider the optimal control problem: Minimize

$$E\{\int_0^T ((M(t)x(t),x(t)) + (N(t)x(t),u(t)))dt + P_0x(T),x(T))\} \tag{6.1}$$

over all $u \in M_W^2(0,T;U)$ subject to

$$dx = (Ax + Bu)dt + (C(x) + D(u))dW_t$$
$$x(0) = x_0 \tag{6.2}$$

Here we assume

(a) H,K,U *are real Hilbert spaces;*
(b) W *is a* K-*valued Brownian motion with covariance operator* S*;*
(c) A *is the infinitesimal generator of a* C_0-*semigroup such that*

$$|e^{tA}| \leq \mu e^{\omega t} \qquad \mu > 0; \tag{6.3}$$

(d) $B \in \mathcal{L}(U,H)$, $C \in \mathcal{L}(H;\mathcal{L}(K,H))$, $D \in \mathcal{L}(U,\mathcal{L}(K,H))$; and
(e) $M(t) \in \Sigma^+(H)$, $N(t) \in \Sigma^+(U)$, $N(t) \geq \varepsilon \geq 0$, $P_0 \in \Sigma^+(H)$.

We use here the notation of §8 in Chapter 2.

We remark that, under hypotheses (a)-(e), the Cauchy Problem (6.2) has a unique mild solution (see Chapter 1).

The Riccati equation associated with Problem (6.1) is

$$P'(t) - A^*P(t) - P(t)A + (P(t)B + \Pi(P(t))(N(t) + \Delta(P(t)))^{-1}$$
$$(B^*P(t) + \Pi^*(P(t))) - \Lambda(P(t)) = M(t) \tag{6.4}$$
$$P(0) = P_0$$

where the linear operators Δ,Π,Λ are defined in the following way:

$$(\Delta(T)\alpha,\beta)_U = tr(D(\alpha)SD^*(\beta)T), \ \forall\alpha,\beta \in U, \ \forall T \in \mathcal{L}(H) \tag{6.5}$$

$$(\Pi(T)\alpha,\beta)_U = \frac{1}{2} tr(C(\alpha)SD^*(\beta)T + D(\beta)SC^*(\alpha)T) \tag{6.6}$$
$$\forall\alpha \in H, \ \beta \in U, \ T \in \mathcal{L}(H)$$

$$(\Lambda(T)\alpha,\beta)_H = \mathrm{tr}(C(\alpha)SC^*(\beta)T); \quad \forall\alpha,\ \beta \in H,\ T \in \mathcal{L}(H). \tag{6.7}$$

We have $\Delta \in \mathcal{L}(\mathcal{L}(H), \mathcal{L}(U))$, $\Pi \in \mathcal{L}(\mathcal{L}(H); \mathcal{L}(H,U))$, and $\Lambda \in \mathcal{L}(\mathcal{L}(H))$; moreover, the following implication holds:

$$T \in \Sigma^+(H) \Rightarrow \Delta(T) \in \Sigma^+(U) \tag{6.8}$$

and therefore the inverse of $(N(t) + \Delta(P(t)))$ exists if $P(t) \geqslant 0$. As in §8 of Chapter 2, by the 'mild' solution of Problem (6.4) we mean an application $P \in C_s([0,T];\Sigma(H))$ such that

$$P(t)x = e^{tA^*}P_0e^{tA}x + \int_0^t e^{(t-s)A^*}(-P(s)B + \Pi(P(s))$$

$$(N(s) + \Delta(P(s))^{-1}(B^*P(s) + \Pi^*(P(s)) + \Lambda(P(s)) + M(s)) \tag{6.9}$$

$$e^{(t-s)A}\ x\ ds,\ \forall x \in H.$$

Along with (6.4), we consider the approximating problem

$$P_n' - A^*P_n - P_nA + (P_nB + \Pi(P_n))(N + \Delta(P_n))^{-1} \tag{6.10}$$

$$(B^*P_n + \Pi^*P_n) - \Lambda(P_n) = M.$$

We need the following lemma:

LEMMA 6 *Let* $\gamma \in \mathcal{L}(\mathcal{L}(H))$, $S_0 \in \Sigma^+(H)$, $\lambda \in \mathbb{R}$, $F \in C_s([0,T];\Sigma(H))$ *with* $F(t) \in \Sigma^+(H)$. *Assume that:*

$$T \in \Sigma(H) \Rightarrow \gamma(H) \in \Sigma(H) \tag{6.11}$$

and let S *be the solution to the Cauchy problem*

$$S'(t) = \gamma(S(t)) + \lambda S(t) + F(t),\ S(0) = S_0. \tag{6.12}$$

Then we have $S(t) \in \Sigma^+(H)$ *for any* $t \geqslant 0$.

Proof Clearly it suffices to prove (6.11) for a fixed λ. Now if $\lambda > \|\gamma\|$ we have

$$\gamma(T) + \lambda T + F(t) \geqslant 0,\quad \forall T \geqslant 0 \tag{6.13}$$

and the conclusion follows by successive approximations.

THEOREM 3 *Under assumptions (a)-(e) there exists a unique solution* P *(resp.* P_n*) to Eq. (6.9) (resp. Eq. (6.10)) belonging to* $C_s([0,T];\Sigma^+(H))$. *Moreover,* $P_n(\cdot)x \to P(\cdot)x$ *in* $C([0,T];H)$ $\forall x \in H$

Proof As in Theorem 10 in Chapter 2 we prove first that there exists a maximal solution P (resp. P_n) to Eq. (6.9) (resp. Eq. (6.10)).

We prove now that $P_n(t) \geqslant 0$; we have in fact

$$P_n' = \gamma_n(P_n) + M \tag{6.14}$$

where

$$\gamma_n(T) = A_n^*T + TA_n - \frac{1}{2}(TB + \Pi(T))(N + \Delta(P_n))^{-1} \tag{6.15}$$

$$(B^*P_n + \Pi^*(P_n)) - \frac{1}{2}(P_nB + \Pi(P_n))(N + \Delta(P_n))^{-1}$$

$$(B^*T + \Pi(T)) + \Lambda(t).$$

By (6.8), we see that γ fulfils (6.11), so by Lemma 6 $P_n(t) \geqslant 0$ for all t in the maximal interval of existence of P_n.

Now by (6.9) (with $A = A_n$, $P = P_n$) we get

$$(P_n(t)x,x) \leqslant (e^{tA}P_0x, e^{tA}P_0x) \tag{6.16}$$

$$+ \int_0^t (M(s) + \Lambda(P_n(s))e^{(t-s)A}x,\ e^{(t-sA}x))\,ds$$

$$|P_n(t)| \leqslant \mu^2 e^{2|\omega|t}|P_0| + \int_0^t \mu^2 e^{2|\omega|(t-s)} \tag{6.17}$$

$$(|M(s)| + |\Lambda|\ |P_n(s)|)ds$$

and by Gronwall's lemma the *a priori* estimate $|P_n(t)| \leqslant$ cost follows. So the solution to Eq. (6.10) is global and, arguing as in Theorem 10 of Chapter 2, we find $P_n(t)x \to P(t)x$ $\forall x \in H$ where P is the solution to Eq. (6.9).

REMARK 3 Propositions 4 and 5 of Chapter 2 on the regularity of solutions can be easily generalized.

We give, finally, an application to the control Problem (6.1).

We set

$$\psi(t,x) = \frac{1}{2}(P(T-t)x,x), \quad \psi^n(t,x) = \frac{1}{2}(P_n(T-t)x,x). \tag{6.18}$$

Arguing as in the proof of Proposition 3 of Chapter 2 we see that ψ and ψ^n fulfil, respectively, the equations:

$$\begin{aligned}&\psi_t(t,x) - \frac{1}{2}((N(t) + \Delta(\psi_{xx}(t,x))^{-1}(B^*\psi_x(t,x)\\&+ \Pi^*(\psi_{xx}(t,x))x), (B^*\psi_x(t,x) + \Pi^*(\psi_{xx}(t,x)x)))\\&+ (Ax,\psi_x(t,x)) + ((\Lambda(\psi_{xx}(t,x))x,x) + \frac{1}{2}(M(t)x,x) = 0\end{aligned} \tag{6.19}$$

$$\begin{aligned}&\psi_t^n(t,x) - \frac{1}{2}((N(t) + \Delta(\psi_{xx}^n(t,x))^{-1}(B^*\psi_x^n(t,x)\\&+ \Pi^*(\psi_{xx}^n(t,x))x), B^*\psi_x^n(t,x) + \Pi^*(\psi_{xx}^n(t,x))x)\\&+ (A_nx,\psi_x^n(t,x)) + \frac{1}{2}(M(t)x,x) + ((\Lambda\ \psi_{xx}^n(t,x)x,x) = 0\end{aligned} \tag{6.20}$$

with

$$\psi(T,x) = \psi^n(T,x) = \frac{1}{2}(P_0x,x). \tag{6.21}$$

LEMMA 7 *Assume that hypotheses (a)-(e) hold, and let* $u \in M_W^2(t,T;U)$, $x \in H$ *and* y *be a solution of the Cauchy problem*

$$\begin{aligned}&dy(s) = (Ay + Bu)ds + (C(y) + D(u))dW_s\\&y(t) = x.\end{aligned} \tag{6.22}$$

Then, the following identity holds:

$$\begin{aligned}&\psi(t,x) + \frac{1}{2} E \int_t^T |(N + \Delta(\psi_{xx}))^{-1/2}(B^*\psi_x(s,y)\\&+ \Pi^*(\psi_{xx}(s,y)y) + (N + \Delta(\psi_{xx}))^{1/2}u(s)|^2 ds\\&= \frac{1}{2} E \{\int_t^T (\frac{1}{2}(M(s)y(s),y(s)) + \frac{1}{2}(N(s)u(s),u(s))ds\\&+ \frac{1}{2} (P_0y(T),y(T))\}.\end{aligned} \tag{6.23}$$

Proof We prove (6.23) for $A = A_n$; once this is done, it is not difficult to take the limit as $n \to \infty$ (see Chapter 1). By the Itô formula we have

$$d\psi^n(s,y_n) = (\psi_s^n + (A_n y_n + Bu,\psi_x^n) + \frac{1}{2}(\Delta\psi_{xx}^n(t,y_n)u,u)$$

$$+ (\Pi(\psi_{xx}^n(t,y_n))u,y_n) + \frac{1}{2}(\Lambda(\psi_{xx}^n(t,y_n))y_n,y_n))ds \tag{6.24}$$

$$+ (\psi_x^n,(C(y_n) + D(u))dW_s.$$

Here we have denoted by y_n the solution to (6.22) with $A = A_n$. Now, we recall (6.20), integrate in [t,T], and take the expectation. We thus find (6.23).

The proof of the following proposition is completely analogous (using Lemma 7) to that of Proposition 3, so it will be omitted.

PROPOSITION 4 *Assume that hypotheses (a)-(e) hold, and let* ψ *be defined by (6.18). Then, we have*

$$\psi(t,x) = \frac{1}{2}\min\{E(\int_t^T (M(s)y(s),y(s) + (N(s),u(s)))ds$$

$$+ \frac{1}{2}(P_0 y(T),y(T))); \; dy = (Ay + Bu)ds + (C(y) + D(u))dW_s,$$

$$y(t) = x,\; s \in [t,T],\; u \in M_W^2(t,T;U)\}. \tag{6.25}$$

Moreover, Problem (6.1) has a unique solution u^* *and* u^* *is given by:*

$$u^*(s) = -(N(s) + \Delta(P(T-s)))^{-1}(B^*P(T-s) + \Pi^*(P(T-s)))y^*(s). \tag{6.26}$$

where y^* *is the solution of the closed loop differential equation*

$$dy = (Ay - B(N + \Delta(P(T-s)))^{-1}(B^*P(T-s)$$

$$+ \Pi^*(P(T-s)y)ds + (C(y) - D((N + \Delta(P(T-s))^{-1}(B^*P(T-s) \tag{6.27}$$

$$+ \Pi^*(P(T-s))y)dW_s$$

$$y(0) = x.$$

§7 References

[1] A. Bensoussan, Variational methods in stochastic control, Corso CIME, Springer, Berlin, (1981).

[2] A. Bensoussan, J.L. Lions, *Temps d'arrêt et contrôle impulsionnel*, Dunod, Paris, (1978).

[3] V. Barbu, G. Da Prato, A direct method for studying the dynamic programming equation for controlled diffusion processes in Hilbert spaces, *Numer. Funct. Anal. and Optimiz.* 4 (1), 23-43 (1981).

[4] V. Barbu, G. Da Prato, Solution of Bellman equation associated with an infinite dimensional stochastic control problem and synthesis of optimal control, to appear in *SIAM J. in Control and Optimization.*

[5] R.F. Curtain, Linear stochastic Itô equations in Hilbert spaces, *Stochastic differential systems*, Springer, Berlin, (1980).

[6] R.F. Curtain, A.J. Pritchard, *Infinite dimensional linear systems theory*, Springer, Berlin, (1978).

[7] G. Da Prato, M. Iannelli, L. Tubaro, Semilinear stochastic differential equations, *Boll, UMI* (1978) 168-185.

[8] M.H.A. Davis, Martingale methods in stochastic control, *Stochastic differential systems*, Springer, Berlin, (1980).

[9] W.H. Fleming, R.W. Rishel, *Deterministic and stochastic optimal control*, Springer, Berlin, (1975).

[10] A. Friedman, *Stochastic differential equations and applications*, Academic Press, New York, (1975).

[11] A. Ichikawa, Dynamic programming approach to stochastic evolution equations, *SIAM J. in Control and Optimization* 17, (1), (1979) 152-174.

[12] P.L. Lions, Equations de Hamilton-Jacobi-Bellman dégénérées, *Comptes-Rendus*, 289, Paris, (1979) 329-332.

[13] H.J. Kushner, *Probability methods for approximation in stochastic control and for elliptic equations*, Academic Press, New York, (1980).

[14] M. Metivier, T. Pellaumail, *Stochastic integral*, Academic Press, New York, (1980).

[15] E. Pardoux, Intégrales stochastiques Hilbertiennes, *Cahiers de Mathématiques de la Decision*, Université Paris-Dauphine, (1971).